THE DIMENSIONS OF PARADISE

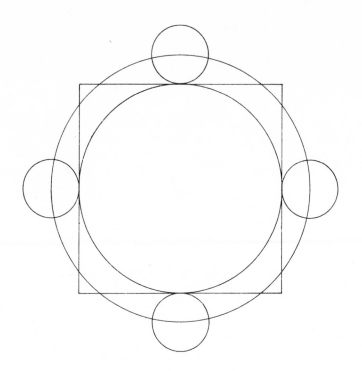

THE DIMENSIONS
OF PARADISE

The proportions and symbolic
numbers of ancient
cosmology

JOHN MICHELL

1817

Harper & Row, Publishers, San Francisco

Cambridge, Hagerstown, New York, Philadelphia, Washington
London, Mexico City, São Paulo, Singapore, Sydney

This book summarizes and extends researches first published in
the author's *City of Revelation*, Garnstone Press, 1971

FIRST U.S. EDITION

LC: 87-46220

88 89 90 91 92 10 9 8 7 6 5 4 3 2 1

CONTENTS

Summary: in quest of the canon

ANCIENT SCIENCE was based like that of today on number, but whereas number is now used in the quantitative sense for secular purposes, the ancients regarded numbers as symbols of the universe, finding parallels between the inherent structure of number and all types of form and motion. Theirs was a very different view of the world from that which now obtains. They inhabited a living universe, a creature of divine fabrication, designed in accordance with reason and thus to some extent comprehensible by the human mind.

The special regard paid to mathematical studies in the ancient world arose from the understanding that number is the mean term in the progression from divine reason to its imperfect reflection in humanity. At some very early period, by a process quite beyond explanation, certain groups of numbers were brought together and codified. Thus was created that numerical standard, or canon of proportion, which was at the root of all ancient cultures and was everywhere attributed to some form of miraculous revelation. It was taken to be the nucleus and activating principle of number generally, a summary of all the types of progressions and relationships which occur within the field of number and thus a faithful image of the numerically created universe.

In the known civilizations of antiquity, as China, Babylon and Egypt, the canon of number was venerated as the source of all knowledge and a guide to rightful conduct. Its influence extended from art and music to affairs of state. Every branch of science expressed its theories and observations in terms of that same small group of numbers which is investigated in this book. One numerical code has fashioned the whole of ancient mathematics, music, astronomy, chronology, metrology and every variety of craft. It has left its mark on every relic and tradition of ancient cultures. There is nothing artificial about it, for the conclusion to these researches is that the various orders of natural phenomena do indeed conform to certain similar patterns of number, which also provide the framework of number itself. This allows the eventual reconstitution of that scientific standard which supported the fabric of ancient societies over periods of time which, by modern reckoning, seem remarkably long.

The author is frankly partial to the traditional order of philosophy, associated in the West with Plato but also expressed in every different culture. It is called idealistic because it is concerned with causes rather than effects and with ideal forms rather than appearances. It is also called perennial, meaning that it grows naturally in the human mind and blossoms at certain seasons. The reason for its constant, universal recurrence is that it is mathematically based. Thus it provides a most realistic view of the world, balanced and made fully human by its transcendental aspect, the traditional doctrine of the soul's immortality.

The number series which is demonstrated throughout this book not only was the source of all traditional arts and sciences but also gave birth to the system of philosophy adopted by Plato. That philosophy occurs of its own accord in the mind of whoever studies these numbers, their relationships and their applications in different branches of nature. Through such studies the Pythagorean dictum, All is Number, comes to life, and thus is opened a new outlook on the world in general. This new outlook, which is not in fact new but traditional, has certain consequences, discussed in the final chapter. It opens possibilities for the future, when the lack of a common, humanistic, scientific standard in affairs has become even more glaring than it is now, and necessity compels the search for some universal guiding principle.

To that future this book is dedicated. Its contents may in part seem dense and obscure. That is mainly due to the author's lack of ability to do justice to his worthiest of all subjects. But it is due also to the modern eclipse of the traditional world-view which gives significance to these studies. The science here unfolded is of no obvious relevance to the modern world, and the type of philosophy which goes with it has been supplanted by other, temporary orthodoxies. To give sense and context to the following studies in the ancient canon of number, the traditional sciences relating to it are briefly described in separate chapters, with reference to the grand alchemical science to which they all contributed and the cosmological outlook which engendered them. This last is the most important, for the purpose and methods of the old sciences are only apparent in the light of traditional cosmology. That light is still hidden below the horizon of modern consciousness; but it can never be extinguished. And when next it arises, demanding the forms of science appropriate to it, the subjects raised in this book will once again be of paramount importance.

This book is the outcome of its author's quest, pursued over many years, for the legendary key to universal knowledge alluded to in esoteric traditions and early texts. Its point of departure was Plato's statement in the *Laws* that the Egyptian priests possessed a canon of lawful proportions and harmonies, by means of which their civilized standards had been preserved uncorrupted for literally thousands of years. The discovery and maintenance of true cultural standards was the main theme of Plato's own writings. His scheme for a well-governed city, described in his *Laws,* was based on a certain numerical formula, often referred to but specified by only one of its components, the number 5040. From this and his other mathematical allusions the inference is that Plato himself had studied the laws of harmony he attributed to the Egyptians. In pure mathematical form those laws were made the cornerstone of his proposed reforms in education and politics. The following chapters on Platonic number show how the laws of harmony were expressed numerically, as the dimensions of a city, a scale of music or intervals in astronomy. In all his cosmological demonstrations Plato used the same set of numbers and similar geometrical diagrams, applying them to such apparently different things as music and the order of the planets, and thus illustrating his belief that number is the 'natural bond' which holds together the entire universe.

One of the conclusions from this study is that Plato's symbolic arithmetic was not a contemporary discovery but a heritage from the distant past. The groundplan of his imaginary City (figure 47) consists of the same combined shapes and numbers as the Stonehenge plan (figure 6), laid down some 1500 years earlier. Their common units of measure were derived from the same archetype, the numerical image of the cosmos. In Chapter 3 the ancient units are analysed and given their exact values, from which it appears that their lengths represent subdivisions of certain basic standards. The standards referred to are not in the first instance physical: they are indeed reflected in the actual dimensions of the earth and the solar system, but in essence they are purely numerical. And the numbers which express ancient units of length are the same as those which denote the scales of traditional music. The forms of music and measure known to Plato were defined and codified thousands of years before his time. Their common source was the canon of number which Plato either learnt wholly from certain teachers or partly reconstructed. His own concern as a would-be reformer and cultural revivalist was to renew the influence of the canon and make it once more effective as an instrument for universal harmony and stability.

9

Mystery schools

It was probably through the Mystery schools, the select institutions of scholarship and mystical inquiry in classical and early Christian times, that the esoteric tradition which Plato drew on was passed down to the founders of Christianity. They were the cause of its last flourishing, soon to be blighted. St John's New Jerusalem, a visionary form of Plato's ideal City and numerically identical to it, was a token of the 'new heaven and new earth' which the prophet foresaw as the issue of renaissance through Christianity and the restoration of the true cosmic standard. Early Christian traditions are particularly useful through indicating the relative meanings and importance attributed to those numbers which occur in earlier sacred contexts. Thus for instance we learn of the supreme significance to Christian mystics of the number 3168, which was also the paramount number in the ancient canon. Those Christian sects who practised the numerical theology claimed that, through assimilating the pagan science, Christianity had become the legitimate heir to the ancient religious tradition.

3168

The significance of this present subject can be summed up in many different ways. To artists, architects and musicians the study of number and proportion has been of traditional interest, and when the current vogue for novelty and individualism has run its course, it will become so again. With scientists of all disciplines the case is similar. Bereft of guidance by any common philosophy, their researches and products are determined by the whims of commerce, militarism, national pride and similar vanities; and the world of scholarship is likewise dominated by faddish intellectualism. Thus are created weird, aberrant thought-forms and monstrous manifestations. At such periods of philosophical anarchy, says Plato in the *Republic*, when there is no common means of distinguishing between beneficial and destructive products, popular demand arises for a standard of judgment. This is usually answered by some tyrant with his own prescription for standards. The demand, of course, is for an objective standard, one that is rooted in nature and reflects no particular theory or ideology. With this consideration begins the quest for the venerable cosmic standard or unified world-image which is numerically structured to represent in essence the entire universe.

cosmic standard

1 The heavenly city as eternal standard

THE PRESENT AGE is commonly perceived as a time of crisis in which all civilized institutions are threatened by unpredictable forces and the future of life on earth is by no means assured. There is evidence enough to justify such forebodings, but popular belief in the imminent destruction of the world has not been confined to our own period: it has recurred throughout history, and earlier legends tell of universal floods, fires and periodic cataclysms in the distant past.

At such times a certain form becomes activated within the contemporary mind, the image of a celestial city. Sightings of the aerial New Jerusalem, as described by St John the Divine in Revelation at the beginning of the Christian era, have been reported on many subsequent occasions, often at moments of millenarian excitement. Many Christians early in the third century believed that they would live to witness the descent of the New Jerusalem, and their hopes were raised when a beautiful walled city became apparent in the skies over Judaea. According to Tertullian, it was seen every morning for forty days, fading away as the dawn lightened. Some eight hundred years later, bands of poor pilgrims, struggling across Europe towards Jerusalem during the People's Crusade, were sustained by visions of a glorious city in the air above them to which ghostly crowds were flocking.

The traditional interpretation of such visions is that they presage the coming of a new dispensation which will reproduce on earth the harmonious order of the heavens. In the symbolism of all religions a geometric construction representing the heavenly city or map of paradise has a central place. It occurs in sacred art as a mandala, a concentric arrangement of circles, squares and polygons, depicting in essence the entire universe. Related images include the labyrinth, the paradisial garden, the walled enclosure or temple precinct, the world tree, the enchanted castle on a rock, the sacred mountain, stone, well or spring and all other symbols of the universal axis which remains fixed and constant amid the ever-changing world of phenomena. The effect of these symbols is to exert an orderly influence on society and the individual mind. C.G. Jung has written much about the therapeutic powers of mandala patterns and their occurrence in periods of mental disturbance.

The association between the ideal city image and states of collective and individual madness has caused psychologists and historians generally to emphasize its function as a compensatory fantasy, occurring to oppressed, fanatical or feeble-minded people in times of trouble. Yet its reality as an *archetype*, an innate component of human mentality, is apparent from its spontaneous appearance in different ages, and its potential is by no means exhausted by its effect on the morbid. The ancient philosophers venerated the established image of the celestial city and based all their studies on it, regarding it as the true, revealed image of God's creation and thus the appointed standard for all human affairs. In terms of the 'sacred' units of measure (those which represent fractions of both the universe as macrocosm and the human microcosm), its dimensions displayed certain numbers, which were also prominent in ancient astronomy, time-keeping and all other numerical sciences.

These same numbers were also found to express the harmonic intervals in the musical scale. Long before Pythagoras made his famous experiments with lengths of string and pipe, the relationship between number and sound had been noted, and ancient rulers specified certain lawful scales which had to be followed in all musical compositions. The reason for this was that they recognized music as the most influential of all arts, appealing directly to the human temper, and thus a potential cause of disturbance in their carefully ordered canonical societies.

The canon of number, encodifying the dimensions of the world, sacred measures and the ratios of geometry and music, was preserved in the state temples which were the centres of ancient government and education. There it was studied by all who aspired to practise priestcraft, politics or any of the arts which influence the forms of society. By this means, said Plato, civilizations governed by the canon were kept stable and uncorrupted over thousands of years. In his time, the fourth century BC, the Egyptians alone maintained their state canon. He refers to it in Book II of the *Laws* (656).

The important passage where the canon is mentioned is part of a conversation between a Cretan, Clinias, and his visitor, a senior political philosopher from Athens whose ideas are closely identified with Plato's. He is speaking about the influence of music and dance forms on the characters of young people who practise or are exposed to them. It surprises him that modern governments allow composers to publish whatever tunes and lyrics they fancy, without regard for the consequences. The only exception to this lax state of affairs, he says, is Egypt:

'It appears that long ago it was decided by that nation to follow the rule of which we are speaking, that the movements and melodies executed by the young generation should be intrinsically good. The types to be permitted were defined in detail and were posted up in their temples. Painters and all other designers of dance or imagery were forbidden to deviate from these types or introduce new ones, either in their own arts or in music. The traditional forms were, and still are, maintained. If you go to look at their art you will find that the works of ten thousand years ago – I mean that literally – are no better or worse than those of today, because both ancient and modern were based on the same standards.'

The question then arises of who first instituted the canon of proportion in art and music, and the Athenian continues:

'One could doubtless find things to criticize in other Egyptian institutions, but in the matter of music it is a very noteworthy fact that it has proved possible to canonize those forms of music which are naturally correct and establish them by law. This must originally have been the work of a god or a god-like being, and indeed the Egyptians attribute the forms of music which have been so long preserved to Isis.'

If only we knew the true canonical forms of music, says the Athenian, there could be no valid objection to making them compulsory by law. People who run after pleasure and novelty would no doubt scoff at the traditional music and call it old-fashioned, but such people carry no real weight, and the Egyptians have been able to uphold their canon despite them.

By Plato's time the very idea of a canon of music had been forgotten everywhere except in the academies of Egypt, but he himself had evidently studied and learnt it, for the number code behind it is at the root of all his mathematical allegories and provided the scientific basis of his philosophy. Had he ever succeeded in his ambition to govern a state or become adviser to some enlightened ruler, he would doubtless have restored the canonical system. But the spirit of the age was against him. The wane of theocratic rule in the ancient world and the rise of democracy had removed the sanctions by which the canon was formerly upheld, and its downfall was later made complete by Christianity's opposition to the study of pagan science. The

canon of music [handwritten marginal note]

effects of its declining influence had long been apparent through a corresponding decline in social stability and continuity of government. Ancient Egyptian records tell of primeval dynasties of gods ruling for ten thousand years, followed by patriarchal reigns of from 900 to 150 years and finally by rulers with the present life-spans. In recent times the process has accelerated to the point where modern governments, faced with constant change and uncertainty, survive from day to day by mere expediencies.

In the present time, when the pleasures of democracy are becoming strongly diluted by anxieties about its future, it is natural to be curious about the nature of that canon of proportion to which Plato attributed the long life of ancient civilizations. In calling it 'the work of a god or a god-like being', he implies that it was no mere product of human opinion but a true reflection of the cosmos, discovered or revealed from nature in the distant past and effective in all ages as a permanent standard of reference. Such a device is scarcely dreamed of today, for the modes of thought and philosophy associated with its use are quite alien to those now dominant. But the destructive consequences of modern philosophy are now all too plain, and it may be that times are favourable for the renewal of a traditional way of thought which considers the constant elements of human nature in relation to its natural environment. There are obvious attractions in a philosophy which requires no act of faith for its acceptance but develops from the study of universal laws, encoded in number. It is indeed that 'perennial' philosophy, the periodic recurrences of which are associated with the renewal of civilization and culture. Thomas Taylor, the Platonist, published a resounding advertisement of it in 1791, in his *Dissertation on the Eleusinian and Bacchic Mysteries*:

> As to the philosophy by whose assistance these mysteries are developed, it is coeval with the universe itself; and however its continuity may be broken by opposing systems, it will make its appearance at different periods of time, as long as the sun himself shall continue to illuminate the world. It has, indeed, and may hereafter, be violently assaulted by delusive opinions; but the opposition will be just as imbecil as that of the waves of the sea against a temple built on a rock, which majestically pours them back.

With this encouragement we investigate in the following chapters the form and content of the ancient canon of number and proportion, together with its applications in music, geometry, architecture, astronomy, city and social planning and individual psychology. We are seeking the lost ground-

[handwritten note:] coeval – orginating or existing during the same period of time, lasting through the same era

plan of that ideal city which Plato described as 'a pattern set in the heavens, where those who want to see it can do so, and establish it in their own hearts'.

The cosmic temple

The mythological account of social origins, which naturally appeals to poets, identifies the first human rulers as the Orphic bards, who kept order by music alone, enchanting whole countries with a cycle of songs in harmony with the changing seasons. Plato in *Critias* says that these rulers were divine: 'They did not use physical means of control, like shepherds who direct their flock with blows, but brought their influence to bear on the creature's most sensitive part [which Plato elsewhere states to be susceptibility to music], using persuasion as the steersman uses the helm, to direct the mind and so guide the whole mortal creature.'

Orphic bards

Thereafter, so it is said, arose a caste of priests and law-giving Druids, who reduced the bardic chants to a strict canon and enforced by violence the laws which previously had been gently uttered and spontaneously obeyed. Thus was the gold of primeval times diluted with baser metal, a process which seemingly reflects the change from nomadic life and the reign of free spirit to settled communities under the rule of law.

Druids

Poets, prophets, anarchists and idealists are natural partisans of the golden age and deplore its corruption with the rise of government. There appears, however, to be no planned or rational way in which we can return ourselves to a state of perfect mutual accord and harmony with nature – if ever such a state actually existed. Plato refers to it in Book v of the *Laws* as a state where there are no possessions or marriages, where children are part of the community rather than belonging to individual parents and where people's interests and perceptions are so similar that it is as if they shared even their eyes, ears and hands. This is the ideal which should be in the minds of all designers and administrators of state constitutions. But in practice they must be content with something less, a social order which reflects the ideal and, however imperfectly, aspires to reproduce it.

Plato's modest proposal is for a 'second-best' form of constitution. He claims for it that, if properly founded and maintained, it 'would come very near to immortality and be second only to the ideal'. His practicable scheme for a city, Magnesia, described in the *Laws*, is a model community ruled, as in ancient times, by the canon. The strictness with which Plato defines its laws and the civic duties of its inhabitants has provided an excuse for those who dislike his hard logic to dismiss him as a totalitarian and opponent of liberty. In fact, he was the very opposite of those things. Plato's intention was to free his fellow-citizens from slavery and tyranny by discovering the social and psychological causes of such evils. In the *Republic* is his famous account of how even the best constitutions tend to dissolve into an anarchic form of democracy, which is enjoyable for a time because everyone does exactly what they want, but ultimately becomes intolerable for lack of standards, all distinctions between good and evil having been abolished. The inevitable outcome, says Plato, is that a dictator is acclaimed to power, and he indeed provides standards of a sort, but they are merely to support his own rule. Thus the natural, equitable laws of a canonically structured society are replaced by the arbitrary decrees of a tyrant.

The understanding behind all Plato's essays in ideal constitution-making is that settlements and civilizations are not natural growths but human constructions. Like all artifacts, they are subject to decay, but the process can be retarded and made scarcely perceptible by adopting a constitution designed by specialist craftsmen using the best possible model. The type of craftsmen Plato had in mind were philosophers, and the archetype he required them to follow was the plan of the celestial city, representing the complete order of the universe. Its traditional formula was evidently known to Plato, for he refers to it cryptically in several works, particularly the *Laws*. The unique property of that formula is that it serves to reconcile all the disparate, opposite and complementary elements in creation. In applying it to the foundation of Magnesia, Plato intended to reconcile the opposite principles of freedom and order, forming a society in which a natural, lasting, unobtrusive style of government was combined with individual fulfilment.

The process of distinguishing, separating and then marrying opposite elements is properly the task of the alchemist, and the way of thinking behind all branches of ancient science was essentially alchemical. The following chapters elucidate the basic formula behind Plato's and other examples of the canonical city, but its significance can not be appreciated without some knowledge of the scientific tradition to which it belongs.

Central to that tradition was the temple, which was also the main institution of ancient government. Like all products of the canon, the temple was also a world-image, synthesizing in its proportions the measurements of the human frame with those of the cosmos. It was designed on the principle that 'like attracts like', on the understanding, as Plotinus put it, that if one wishes to attract any aspect of the universal spirit one must create a receptacle in its image. Thus every ancient temple contained symbolic references in its shape, dimensions and furnishings to the god to whom it was dedicated, while the state temple, placed at the conceptual world-centre, represented the entire universe.

Symbolism in the ancient world was always related to practical function, and the symbolic features of the temple were intended to assist the purpose for which the building was designed, as an instrument of invocation. The methods by which it was operated, together with the whole question of ancient sacred technology, are still the deepest of mysteries, but the legends and symbolism of, for example, the temple at Jerusalem indicate that two forms of natural energy were involved, one terrestrial, the other from the atmosphere. Through the ritual fusion of these two elements at certain seasons, a spirit was generated and spread like an enchantment from the temple, bearing fertility to the surrounding countryside and pervading it with an atmosphere of peace and happiness.

temple at Jerusalem [handwritten marginal note]

This subject is expanded in Werner Wolff's scholarly work, *Changing Concepts of the Bible*, which contains some shrewd speculations about the Temple of Solomon as an alchemical generator and storehouse of sacred energy. From the descriptions of the Temple in the Old Testament and in Josephus's *Antiquities of the Jews*, it is shown that its inner chamber, the Holy of Holies, which was lined with gold and contained apparatus of brass and other specified materials, could have functioned in accordance with its legends as an electrical device. Wolff suggests that the power of lightning was harnessed, together with the volcanic energy of the earth, creating the effect known as the Glory of God or the Great Radiance which at certain seasons blazed forth from the Temple.

An earlier writer, Salverte (quoted in Thomas Milner's *Gallery of Nature*), concluded from Josephus's account that 'a forest of spikes with golden or gilt points, and very sharp, covered the roof of this [Jerusalem] temple', and that 'this roof communicated with the caverns in the hill of the temple, by means of metallic tubes, placed in connection with the thick gilding that covered the whole exterior of the building; the points of the spikes there necessarily produce the effect of lightning rods'.

Similar pointed rods appear on the roofs of other temples depicted on ancient coins. In this, and in the many legends which connect sacred sites with lightning activity, is a strong hint that one party in the sacred marriage performed at the Temple was electrical current from the atmosphere.

magnetic Current of earth

The other partner in this union of opposite elements was clearly the magnetic current of the earth, the mercurial, serpentine spirit which corresponds in the life-system of the planet to the subtle energies of the human body, treated by eastern therapists. Temples were sited at nodes and centres of the earth spirit, above fissures and underground caverns where it accumulates. In these catacombs beneath the great temples were celebrated the rites and mysteries of the chthonic deities, which, in terms of spiritual technology, may have involved the fertilizing of the earth spirit by positively charged energy channelled through the lightning rods from the temple roof.

The Old Testament chroniclers tell of life-giving streams which rose beneath the temple and flowed towards the four corners of the earth. They are referred to by Dr Raphael Patai in his book on the Jewish Temple traditions, *Man and Temple*:

> Quite a number of legends tell in an interesting variety of versions about this subterranean network of irrigation canals that issue from underneath the Temple and bring to each country its proper power to grow its particular assortment of fruits. If a tree were planted in the Temple over a spot whence the water-vein issued forth to a certain country, it would grow fruit peculiar to that country; this was known to King Solomon, who accordingly planted in the Temple specimens of fruit trees of the whole earth.
>
> The waters that issued forth from the Temple had the wonderful property of bestowing fertility and health. Legends have it that as in days of old so again in the days of the Messiah 'all the waters of creation' will again spring up from under the threshold of the Temple, will increase and grow mighty as they pour forth all over the land.

The water beneath the Temple was both actual and metaphorical, existing as springs and streams, as spiritual energy and as a symbol of the receptive or lunar aspect of nature. The meaning of that principle is too wide and elusive for it to be given any one name, so in the terminology of ancient science it was given a number, 1080. Its polar opposite, the positive, solar force in the universe, was also referred to as a number, 666. These two numbers, which have an approximate golden section relationship of 1:1.62, were at the root of the alchemical formula which expressed the supreme purpose of the Temple. Not merely was it used to generate energy from fusion of atmospheric and terrestrial currents, but it also served to combine in harmony all the correspondences of those forces on every level of creation. The union of 666 and 1080 is 1746, and when, later on, the arithmetic and

1080
666
golden section relationship
1:1.62
1746

symbolic qualities of that number are investigated, it is seen how worthily 1746 represents the union which was consummated at the Temple, the marriage between heaven and earth.

The site of the Temple at Jerusalem was reputed to be the spot where the waters of the Flood had erupted and where they had receded again into the abyss. The waters were pressed down by a great rock, which was the foundation of the Temple. Patai writes:

> In the middle of the Temple, and constituting the floor of the Holy of Holies, was a huge native rock which was adorned by Jewish legends with the peculiar features of an *Omphalos*, a Navel of the Earth. This rock, called in Hebrew *Ebhen Shetiyyah*, the Stone of Foundation, was the first solid thing created, and was placed by God amidst the as yet boundless fluid of the primeval waters. Legend has it that just as the body of an embryo is built up in its mother's womb from its navel, so God built up the earth concentrically around this stone, the Navel of the Earth. And just as the body of the embryo receives its nourishment from the navel, so the whole earth too receives the waters that nourish it from this Navel.

Like Noah's Ark, where the corresponding pairs in nature were also brought together, the Temple on its rock was afloat on the void, and however well it was built, however apparently firm was the order it represented, sooner or later the waters would again well up and sweep all away. For those same waters which are the source of health, fertility and inspiration are as potentially dangerous as the fiery element on the other side of the alchemical equation. On the mental plane they are located in the depths of the unconscious mind, from which they may rise up to overwhelm the order of sanity, or inner temple, constructed by each individual. This is the periodic destruction by water which, according to the old philosophers, alternates with the fiery cataclysm which is invoked by excessive predominance of the intellect. At the point where the two forces meet is the citadel of the mind, the seat of judgment, where in daily life their rival claims are assessed and balanced, the success with which that is done depending on the quality of one's overall mental pattern and its adequacy in reflecting the essential reality of things.

The most valuable of benefits conferred by the temple institution was in providing a world-image or pattern of cosmology as an ideal model for the construction and maintenance of sanity. Madness is eternal and a gift from the gods, remarks Plato during his discourse on love in *Phaedrus*, whereas sanity is man-made. But since civilized people can not live by inspiration alone, the rational mind should be composed, like the rational city, after the 'pattern set in the heavens'.

Thus human nature and the order of the universe were seen as products of the one archetype, the pattern which the Creator had in mind when he set about his work. On that perception rested the entire fabric of ancient philosophy and science. The Temple was placed intermediate between the two scales, human and cosmic, and the energies it transmitted were two-way; for it was believed not only that the heavens influenced affairs on earth but that the order of human society affected the entire world of nature. Ceremonies throughout the year at the Temple were meant to imitate, and so procure, the fruitful union of all mutually corresponding elements, those above with those below.

With the decline of the old order the enchantment faded and dissolved. Legends of sacred centres across the ancient world tell the same story as the Jews tell about the Temple at Jerusalem: that when it was destroyed the world fell out of balance, and human happiness and culture have ever since been in decline. No longer could the Matrona, spirit of the land of Israel, tryst with her celestial bridegroom at the Temple, for their nuptial chamber, the Holy of Holies, had been defiled. Her melancholy plight is described in the medieval *Zohar Hadash*, quoted by Patai. Every night she wanders round the precincts of the ruined Temple, wailing:

'My couch, my couch, my dwelling-place . . . . In thee came unto me the Lord of the World, my husband, and he would lie in my arms and all that I wished for he would give me. At this hour he used to come unto me, he left his dwelling-place and played betwixt my breasts. My couch, my couch, dost thou not remember how I came to thee rejoicing and happy, and those youth (the Cherubs) came forth to meet me, beating their wings in welcome . . . . From here went forth nourishment unto all the world, and light and blessings to all! I seek for my husband but he is not here. I seek in every place. At this hour my husband used to come unto me and round about him many pious youths, and all the maidens were ready to welcome him. And I would hear from afar the sound of the twin bells tinkling between his feet that I might hear him even before he came unto me. All my maidens would laud the Holy One, blessed be he, then retire each to her place, and we remained in solitude embracing in kisses and love. My husband, my husband, whereto hast thou turned? . . . Thou didst swear that thou wouldst never cease to love me, saying, "If I forget thee, O Jerusalem, let my right hand forget her cunning . . .".'

This haunting scene illustrates in the clearest possible fashion the true meaning of the Temple as the place of union between the energies of earth and cosmos.

The destruction of the Temple is not, however, the end of the story, for part of its legend is set in the future. One day, it is said, the Temple will be restored, the sacred world-order will again be established and harmony between men and nature will once more prevail. That event, according to all prophecies, will take place at a period of extreme need and desperation. Its timing is beyond human control, but the human mind is presumably the medium through which its coming is made apparent. It may be that Nature, that great, enigmatic organism, directs in a subtle way the intelligence of her human products, re-ordering patterns of mind in accordance with the necessity of the situation. The present interests of Nature, it is now very clear, demand a fundamental change in the prevailing cosmology – our beliefs about the nature of the world and how best to relate to it. That idea sets the mind in pursuit of an ideal cosmology, and on the 'seek and ye shall find' principle, such a quest invokes its object. Awaiting the seeker is the outline of a celestial city, the very same as appeared to St John on Patmos, agreeing in its dimensions with the cosmological cities and temples which sanctified the earth over long ages in the distant past.

The New Jerusalem

The vision of the Holy City which occurred to St John on the island of Patmos is described in Revelation, 21.

> And I saw a new heaven and a new earth: for the first heaven and the first earth were passed away; and there was no more sea.
> And I John saw the holy city, new Jerusalem, coming down from God out of heaven, prepared as a bride adorned for her husband.

A few verses later the account is amplified with details of the City:

> And there came unto me one of the seven angels which had the seven vials full of the seven last plagues, and talked with me, saying, Come hither, I will show thee the bride, the Lamb's wife.

And he carried me away in the spirit to a great and high mountain, and shewed me that great city, the holy Jerusalem, descending out of heaven from God,

Having the glory of God: and her light was like unto a stone most precious, even like a jasper stone, clear as crystal;

And had a wall great and high, and had twelve gates, and at the gates twelve angels, and names written thereon, which are the names of the twelve tribes of the children of Israel:

On the east three gates; on the north three gates; on the south three gates; and on the west three gates.

And the wall of the city had twelve foundations, and in them the names of the twelve apostles of the Lamb.

The twelve-fold division of the City is then further emphasized.

And the foundations of the wall of the city were garnished with all manner of precious stones. The first foundation was jasper; the second, sapphire; the third, a chalcedony; the fourth, an emerald;

The fifth, sardonyx; the sixth, sardius; the seventh, chrysolite; the eighth, beryl; the ninth, a topaz; the tenth, chrysoprasus; the eleventh, a jacynth; the twelfth, an amethyst.

And the twelve gates were twelve pearls; every several gate was of one pearl: and the street of the city was pure gold, as it were transparent glass.

This wonderful vision of a glittering, translucent city is the climax of St John's Revelation. Preceding it are scenes of horror and destruction mingled with rapturous glimpses of paradise. Every element in the drama is contrasted with its opposite, the Whore of Babylon and the Beast from the abyss playing alternate parts with the Bride of Jerusalem and the Lamb on Mount Zion. Its images and processes correspond to those inherent in the mind and allow it to be interpreted on different levels, either as a pattern of historical events or as a course of mental development. According to esoteric tradition, Revelation describes allegorically the unfolding of the ancient Mystery teaching and the initiation of St John himself.

The records of folklore and mystical experience contain many accounts of an ethereal city viewed in the sky, variously explained as an effect of light on clouds, a mirage reflecting an actual city on earth or a vision of paradise. In Celtic lands where, perhaps because of the misty atmosphere, such phenomena are most commonly reported, the ghostly city in the air is said to be the habitation of the saintly dead. It has been claimed that St John was a Celt; but whether or not he ever personally saw such a thing, his poetic instinct caused him to adopt the legendary city in the sky as the external image of that cosmological city, or ideal order of the universe, revealed to him in the course of his studies. The traditional code of cosmology, for which Plato's symbol was a political city-state, is represented by St John in the form of a familiar type of vision.

As proof that the New Jerusalem was no mere illusion or imaginative fancy, St John gives its actual measured dimensions.

And he that talked with me had a golden reed to measure the city, and the gates thereof, and the wall thereof.

And the city lieth foursquare, and the length is as large as the breadth; and he measured the city with the reed, twelve thousand furlongs. The length and the breadth and the height of it are equal.

And he measured the wall thereof, an hundred and forty and four cubits, according to the measure of a man, that is, of the angel.

New Jerusalem a cube —

Evidently the New Jerusalem took the form of a cube with each of its 12 sides measuring 12 000 furlongs, each of its 6 faces 144 million sq. furlongs, and with a volume of 1728×10^9 cubic furlongs. Somehow associated with it is a wall of 144 cubits. However the foot is defined, a furlong is 660 feet and a cubit is $1\frac{1}{2}$ feet, so the two measures are on different scales. Several modern translators of the New Testament have made the problem of reconciling them quite impossible by corrupting the original text. In one version the wall is said to be 144 cubits in height, which St John neither states nor implies, while others are so absurd as to give the dimensions of the New Jerusalem in metres! Reference to the metre in this context are particularly crass, for it is a unit of recent, secular contrivance, designed in the spirit of 'scientific atheism' to bear no relation to any human measure or standard. In contrast to the metric system, the New Jerusalem is not of human manufacture, and the units of measure which pertain to its dimensions are properly called sacred because they derive from eternal standards in number and nature.

In prophetic allusions to the Holy City it is repeatedly emphasized that its dimensions should be measured, using the correct yardstick, the 'perfect and just measure'. Two such injunctions occur in the Old Testament:

I lifted up my eyes again, and looked, and behold a man with a measuring line in his hand. Then said I, Whither goest thou? And he said unto me, To measure Jerusalem, to see what is the breadth thereof, and what is the length thereof. – Zechariah 2,2.

Thou son of man, show the house to the house of Israel, that they may be ashamed of their iniquities: and let them measure the pattern. – Ezekiel 43,10.

St John in Revelation 11 is told, 'Rise and measure the temple of God, and the altar, and them that worship therein.' For that purpose the angel gave him a 'reed like a rod', in Greek καλαμος ὁμοιος ῥαβδῳ. The total numerical value of the letters in that phrase is 1729, which by the conventions of gematria (explained in the following chapter) is equivalent to 1728 or 12^3, and 1.728 ft. is the length of the Egyptian royal cubit (equal to $1\frac{1}{2}$ Egyptian feet) by which the New Jerusalem is measured.

The difference in scale between the 12000 furlongs and the 144 cubits of the New Jerusalem indicates that it represents both the macrocosm and the microcosm, the order of the heavens and the constitution of human nature. Both are measured by the sacred units which apply to the astronomical as well as to the human scale and thus unite the two. When the dimensions of the New Jerusalem are made commensurable as 12 furlongs and 14000 cubits the geometric groundplan of the City becomes visible, for

$$14400 \text{ cubits} = 14400 \times 1.728 = 24883.2 \text{ ft., and}$$
$$12 \text{ furlongs} = 12 \times 660 = 7920 \text{ ft.}$$

The significance of these measures is that a circle of diameter 7920 ft. has a circumference of 24883.2 ft. In ancient sacred or cosmological arithmetic the π ratio between the diameter and circumference of a circle was made rational as a simple fraction, the most convenient being 22/7. For numerical reasons others were also used (see page 66), including 864/275. This is slightly more accurate than 22/7 and it makes the above calculation exact, for $7920 \times 864/275 = 24883.2$.

Thus the basic plan of the New Jerusalem is a square of 12 furlongs containing a circle of circumference 14400 cubits (figure 1).

Measured in miles instead of feet, the wall of the New Jerusalem forms a plan of the earth, for the earth's mean diameter is some 7920 miles and its circumference through the poles is 24883.2 miles. The significance of this number is that $248832 = 12^5$.

The twelve gates, twelve precious stones and other details of New

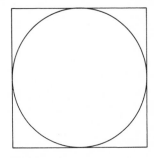
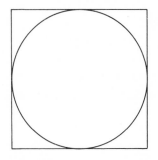

The New Jerusalem:

diameter	= 7920 ft.
circumference	= 24 883.2 ft.
perimeter of square	= 31 680 ft.

The Earth:

diameter	= 7920 miles
circumference	= 24 883.2 miles
perimeter of square	= 31 680 miles

Figure 1.

Jerusalem give clues to the further development of its plan, and the following studies of other cosmological schemes in the same tradition allow the complete diagram to be reconstructed. This in turn opens the way to the rediscovery of the basic number code behind the ideal cities and planetary systems of Plato. We are then led to contemplate the meaning of that mystical city in the heavens whose measures and proportions were held in such ardent reverence by the ancient philosophers.

The twelve hides of Glastonbury

The number twelve which is applied to the furnishings of the New Jerusalem in St John's Revelation, its twelve gates, twelve pearls, twelve foundations, twelve names, twelve jewels and twelve angels, is also the main

Figure 2. Glastonbury Tor.

feature of its dimensions. Its wall measures 120^2 or 14 400 cubits or a tenth part of 12^5 ft. and the 12-furlong square containing it has a perimeter of 6 miles and an area of 144 square furlongs, which is equivalent to 1440 acres (an acre contains 43 560 square feet).

This area of 1440 acres is referred to in the foundation legend of Glastonbury in Somerset. The town with its great ruined Abbey is built on the slopes of a group of hills, the highest of which is the remarkable conical Tor, a landmark visible from far over the Somerset plains, from the neighbouring counties of Wiltshire and Dorset and across the Bristol Channel from South Wales. As a source and repository of legends, Glastonbury is distinguished above all other places in England. Episodes

Figure 3. St Mary's Chapel, Glastonbury.

from the Grail cycle, the tales of King Arthur's heroes and other items of prehistoric lore adapted to Christianity are located in its surrounding landscape, and its ancient reputation as a gateway to another world is continually reaffirmed by its natural attraction to mystics.

The most famous of the Glastonbury legends identifies the Abbey as the site of the world's first Christian church, established shortly after the Crucifixion by a company of twelve missionaries, led by St Joseph of Arimathaea, who had been sent over from Gaul by the apostle St Philip. At Glastonbury they were received by the local king, Aviragus, whose territory had not yet fallen under Roman dominance, and he made them a grant of land, amounting to twelve hides, one hide for each member of the party. Within the twelve hides they built a primitive church, said to have been round and constructed in the local style of timber and wattles, and they dwelt in twelve huts or cells drawn up in a circle around it.

After the founders' deaths the church was kept in repair by visiting saints and ascetics, attracted by the natural and historical sanctity of the spot. In the seventh century it was preserved in a wooden casing coated in lead, and it was left intact when King Ine in 704 built the great Abbey church to the east of it. Following the destruction of the Abbey by fire in 1184, the present St Mary Chapel was built on the site of the old church to perpetuate its mystical dimensions.

The traditions of the apostolic foundation of Christianity in Britain are deeply rooted in popular memory, and were repeated by the people of Somerset into modern times. They are supported by the evidence of Tertullian and Origen writing in the second century, and in the sixth century the historian St Gildas declared that Britain was first illuminated by the light of 'Christ the true Sun' in the latter part of the reign of Tiberius Caesar, who died in the year 37. Of Gildas, William of Malmesbury wrote that he was 'a true scholar and by no means an old fool'. William himself, who in the twelfth century studied in the library of Glastonbury Abbey and had access to manuscripts that have since perished, came to believe in the essential truth behind the Glastonbury legends, and it is a fact that at the great Church Councils of the middle ages the English bishops were given precedence before all on account of their acknowledged claim to represent the earliest foundation. It was only after the Reformation, when the Puritans turned against the old traditions and Roman historians found it in their interest to reject the unique, apostolic origins of the English Celtic church, that the peculiar significance of Glastonbury was called in question.

Glastonbury has been described as Britain's only true national shrine, the *omphalos* or Temple of Britain and the English Jerusalem. It was Avalon, the western isle of the dead, a centre of the ancient mysteries, known from the earliest times as a place of supreme natural sanctity. Celtic saints and hermits were attracted there; many ended their days in a lonely Glastonbury cell, contributing their bones to the unique collection of relics, the glory of the medieval abbey. William of Malmesbury wrote, 'The resting place of so many saints is deservedly called a heavenly sanctuary.'

The traditional sanctity of the twelve hides was acknowledged in the Domesday survey of 1084. In accordance with immemorial custom they were neither surveyed nor taxed, and the Abbot of Glastonbury was sovereign within their borders. He ruled and judged the inhabitants of the twelve hides and appointed his own sheriffs and coroners. The King's officers had no authority in his realm. No one knows when and how these privileges arose. They could well have been a survival from prehistoric times, and it has been suggested that the story of St Joseph itself derives from an earlier foundation legend when Glastonbury was a Druid sanctuary.

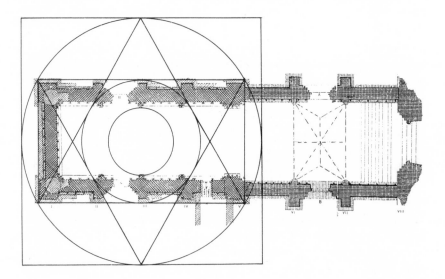

Figure 4. The sacred geometry of St Mary's Chapel, Glastonbury, provided by the New Jerusalem diagram. Its original width is 39.6 ft. and its length, as indicated by the $\sqrt{3}$ rectangle formed by the hexagram, is 68.6 ft. The enclosing square has an area equal to one 10 000th part of 12 hides, or 0.144 acres.

A hide is generally defined as the amount of land which can support a farmer's family and dependents, or as the area worked by one plough in the course of a year. Its area varies with the nature of the land in different parts of the country, but it is most commonly of 120 acres, which is the accepted value in Somerset. Twelve hides are therefore equal to 1440 acres, the same as the area of the 12 × 12 furlong square in the New Jerusalem diagram.

Figure 4 shows a square of 12 hides, but reduced a hundredfold so that its side measures 0.12 furlongs or 79.2 ft. A circle is placed within it as in the New Jerusalem, and a hexagram within that circle contains at its centre a smaller circle whose dimensions are half those of the first. The diameter of the circle within the hexagram is 39.6 ft., and that is also the width of the Glastonbury chapel. Thus the area of twelve hides is linked geometrically with the mystical plan of the original Glastonbury settlement. At the centre of the scheme, contained by the inner walls of the Chapel, is a circle of radius 10.8 ft., representing perhaps the circular church built by St Joseph on that same site.

The New Jerusalem in Stonehenge

Another occurrence of the New Jerusalem diagram in England is in the groundplan of Stonehenge. There is evidence of an ancient connection between that old monument and Glastonbury Abbey, in that the main axis of Glastonbury, from St Benedict's Church, down the length of the Abbey and along Dod Lane towards a former beacon site at Gare Hill in Wiltshire, points at Stonehenge some forty miles to the east. In their respective lore there is only one tradition of a link between the two places. Both are named in the old Welsh Triads as sites of the Perpetual Choirs of Britain, a relic of the Orphic rule that preceded the Druid regime, when the order of society was maintained by music. That subject will be referred to again later. The present demonstration is of the pattern behind the Stonehenge groundplan. In every essential respect it is identical with that of Glastonbury and with the New Jerusalem diagram.

Briefly described, the temple of Stonehenge consists of two concentric stone circles enclosing two U-shaped stone arrangements, all contained within a circular bank and ditch. The following are the dimensions of its various parts as given in R.J.C. Atkinson's *Stonehenge*, in *The Stones of Stonehenge* by the engineer E.H. Stone and in W.M. Flinders Petrie's *Stonehenge Plans*.

Figure 5. Stonehenge restored, with diagram of the outer elevation of the Stonehenge sarsen circle. One lintel is raised to show its curve, the tongued and grooved ends and the mortise holes to receive projecting tenons on uprights.

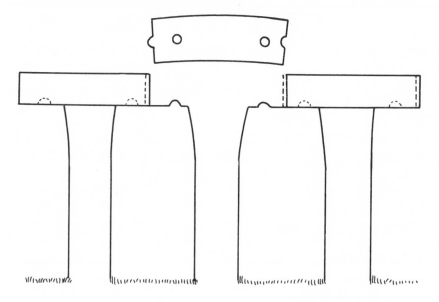

The outer Stonehenge circle consisted of thirty pillars of Wiltshire sarsen stone, of which seventeen only remain, supporting a continuous ring of thirty curved lintel stones, now reduced to six. The lintels were fitted together with tongue-and-groove joints and were held in place by stone tenons protruding from the tops of the pillars to fill holes in the lower surfaces of the lintels. Petrie's careful survey gave 97.32 ft. for the diameter of the circle between the smooth inner faces of its pillars. The width of the lintels is about $3\frac{1}{2}$ ft. (Atkinson), so the outer diameter of the circle is about 104.3 ft. and the mean diameter 100.8 ft.

The mean circumference of the lintel ring is $100.8 \times 22/7 = 316.8$ ft. This measure is confirmed by E.H. Stone, who tackled the problem another way, by measuring the gaps between the upright pillars. The gaps were clearly designed to be half the average width of the pillars themselves. Stone's figure was 3.52 ft., making the width of the pillars 7.04 ft. and the mean circumference of their circle $30 \times (3.52 + 7.04) = 316.8$ ft. It is therefore reasonable to suppose that the intended dimensions of this circle were: mean diameter 100.8 ft.; mean circumference 316.8 ft. or a hundredth part of 6 miles.

Within this circle is another, now ruined and incomplete, formerly consisting of about sixty 'bluestone' pillars, whose origin has been traced by geologists to the Prescelly Mountains in Pembrokeshire. Due to its condition and the inexactness of its lay-out, the authorities are less certain about the dimensions of the bluestone circle than about those of the sarsen ring. Atkinson's estimate of its diameter is 75 ft., Stone's $76\frac{1}{2}$ ft. The width of the bluestones would therefore allow them to be contained in a circle of diameter 79.2 ft. or a hundredth part of the diameter of the circle within the New Jerusalem square.

The central U-shaped arrangement was originally of nineteen bluestone pillars, set in the form of a circle with one end opened up in the direction of midsummer sunrise. Those at the other end, the south-west, 'stand with their inner faces touching the circumference of a semicircle $39\frac{1}{2}$ ft. in diameter' (Atkinson). 39.6 ft. is thus a lawful estimate for the diameter of this semicircle, making it correspond to the inner circle of the New Jerusalem diagram as represented in the Glastonbury Chapel.

Figure 6 shows how perfectly the Stonehenge groundplan correlates with the New Jerusalem diagram superimposed. The square is that of St John's city or the twelve hides of Glastonbury, 7920 ft. wide, and the circle within it is the wall 14400 cubits round, all the dimensions being reproduced on a

Stones now standing ▭   Stones fallen or missing ▭

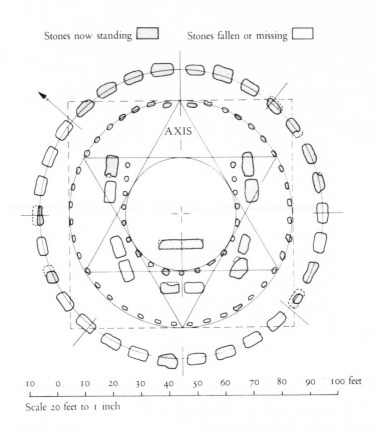

AXIS

10   0   10   20   30   40   50   60   70   80   90   100 feet

Scale 20 feet to 1 inch

Figure 6. The ground plan of Stonehenge overlaid by the New Jerusalem diagram which evidently determined the form of the temple. The mean circumference of the outer, sarsen circle with lintels is 316.8 ft., a hundredth part of 6 miles, and a square with perimeter of 316.8 ft. would contain a circle enclosing the bluestone ring, diameter 79.2 ft. The nineteen stones of the inner U-shaped structure enclose a circle of diameter 39.6 ft.

scale of 1:100. The outer circle has the same perimeter as the square, 316.8 ft. Stonehenge is thus founded on the classic image of sacred geometry, the squared circle representing the reconciliation of opposites, which is the common feature of temples and foundation myths the world over.

The main parts of Stonehenge, those described above, are said to have been built about 4000 years ago, though some of its stones are thought to have been taken from a previous temple on the site, and the outlying parts, such as the bank and the isolated stones near it, also belong to an earlier period. The

Stonehenge builders must therefore have chosen for their model New Jerusalem a site already sacred, and on it they constructed the greatest architectural wonder of the ancient world. The events which prompted them to do so, and the spirit in which they undertook the work, are far beyond our ken, but if the old astrologers' view is correct, and the beginning of each 2160-year 'month' in the Great Year (produced by the movement of the sun's position at the Vernal Equinox through the twelve signs of the sidereal zodiac) marks a period of change and renewal, it may be significant that the great work at Stonehenge took place near the beginning of Aries, while the vision of New Jerusalem occurred to St John near the start of Pisces. The present time, said to be coinciding with the dawn of Aquarius, could thus be auspicious for the revival of that traditional pattern of cosmology which was reputed to provide the most accurate, useful model of human relationship to the cosmos.

The pattern in the heavens

In figure 7 the three versions of the New Jerusalem diagram which have so far been investigated are shown together for comparison, and with them is a fourth. This additional diagram is the astronomical version of the New Jerusalem, representing the 'sublunary world' or the earth and its surrounding atmosphere beneath the influence of the moon.

It is a remarkable fact of nature that the earth and the moon, placed tangent to each other and measured in miles, demonstrate the squared circle in terms of the very same numbers as make up the New Jerusalem diagram.

The earth's mean diameter is 7920 miles and the diameter of the moon is 2160 miles. The combined radii of the two bodies are equal to $3960 + 1080 = 5040$ miles, and a circle with radius 5040 has a circumference (if $\pi = 22/7$) of 31 680 miles. That length, 31 680 or 4×7920 miles, is equal to the perimeter of the square containing the circle of the earth. It is also the measure of the squared circle in the New Jerusalem.

The earth and moon are here in contact with each other, but their actual mean distance apart, about 237 600 miles, is not neglected in the diagram, for if a hexagon is inscribed within the circle of the earth (figure 8) the measure round its six sides is 23 760 miles.

It would be easy to conclude that the New Jerusalem diagram was modelled on this figure of the earth and moon in some remote age when the dimensions of the solar system were reckoned more accurately than was

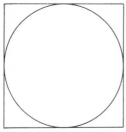

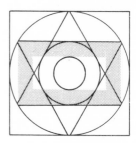

St John's New Jerusalem: diameter, 7920 ft., perimeter of square, 3160 ft.

St Mary's Chapel, Glastonbury: diameter, 79.2 ft., perimeter of square, 316.8 ft., diameter of inner circle, 21.6 ft.

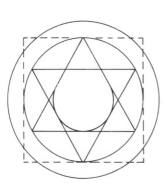

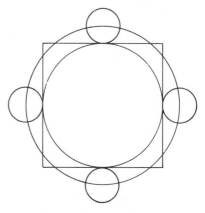

Stonehenge: diameter of (bluestone) circle within square, 79.2 ft., perimeters of outer (sarsen) circle and square, 316.8 ft., diameter of inner horseshoe, 39.6 ft.

Earth and moon: diameter of inner (earth) circle, 7920 miles, perimeters of outer circle and square, 31680 miles, diameter of small (moon) circles, 2160 miles.

Figure 7. Sources of the New Jerusalem diagram.

possible in early historical times. Yet the traditional code of number set out in the New Jerusalem expresses the organization of so many different categories of natural phenomena, including the inherent framework of number itself, that one must be wary in ascribing origins. As one pursues these studies, the world of appearances fades away, its diverse manifestations dissolve into shadows and behind them the mind's eye discerns the reality which sustains their illusions. It is a reality formed by the interplay of creative forces which can most adequately be likened to a pattern of numbers.

The squared circle of earth and moon is therefore best described, not as the

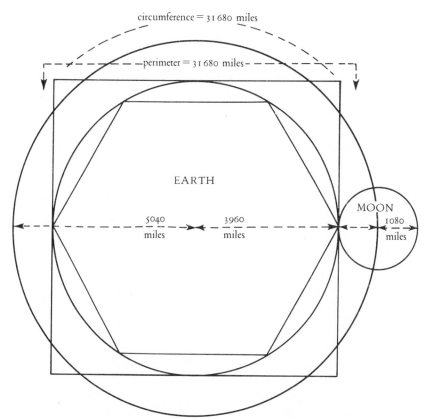

circumference = 31 680 miles

perimeter = 31 680 miles

EARTH

MOON

5040 miles

3960 miles

1080 miles

Figure 8. The squared circle demonstrated by nature in the relative dimensions of the earth and moon. Measured in miles, they supply the numbers of the New Jerusalem diagram. The perimeter of the hexagon, 23 760 miles, is a tenth part of the distance between the earth and the moon.

primary model for the New Jerusalem diagram, but as its astronomical expression.

Constructing the New Jerusalem

The full development of the New Jerusalem plan is illustrated in Figure 9. As the matrix and synthesis of every order of geometry it can be drawn so as to accommodate many other figures, but its framework is firmly tied to its foundation figure, the squared circle. Religious symbols and cosmological patterns the world over, from the mandalas of eastern mysticism to the rose

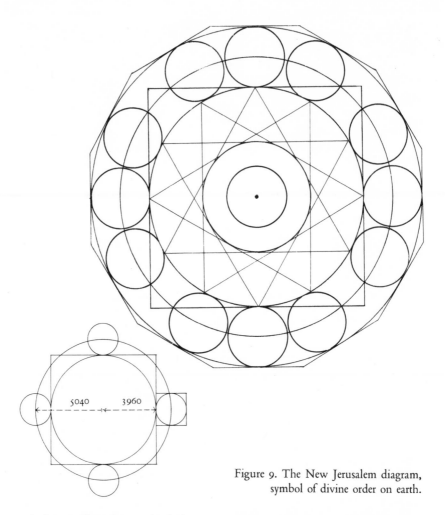

Figure 9. The New Jerusalem diagram,
symbol of divine order on earth.

windows of Gothic cathedrals, are versions of this same diagram. The
earliest example known of it, the plan of Stonehenge, was laid out some four
thousand years ago, but the universality of New Jerusalem imagery and of
the units of measure associated with it suggest an even greater antiquity.

In constructing the New Jerusalem the geometer begins by imitating the
first act described in the Old Testament, 'In the beginning God created the
heaven and the earth.' The corresponding geometric operation is to draw the
circle of the heavens together with the earthly square and to harmonize them
by giving them both an equal perimeter. This is achieved by means of the

Pythagorean 3, 4, 5 triangle as shown on page 69. The sum of these three numbers, $3 + 4 + 5$, is 12, their product, $3 \times 4 \times 5$, is 60, and the numbers 12 and 60 are at the root of the numerical code which supplies all the dimensions of the New Jerusalem. Added together 12 and 60 make 72, and multiplied together they give 720. This is the number of Truth (because $\dot{\eta}$ $\dot{\alpha}\lambda\eta\theta\epsilon\iota\alpha$, the truth, has the value by gematria of 72), and it reveals the truth of things in providing the multiplier which raises the dimensions of the squared-circle figure, created by the 3, 4, 5 triangle, to those of the New Jerusalem. Multiplied by 720 the three sides of the 3, 4, 5 triangle become 2160, 2880 and 3600, the side of the New Jerusalem square becomes 7920,

Figure 10. The underlying geometry of the northern rose window at Chartres extends beyond the window itself (heavy line) as a twelve-fold scheme similar to the New Jerusalem diagram.

the radius of its circle 5040, and both the perimeter of the square and the circumference of the circle measure 31 680.

The circle's diameter divides the squared circle figure into two equal halves, consisting of two half-squares and two semicircles. Their respective areas are consonant with the numerical harmony of the diagram. Measured in units of feet, the area of the half-square is 720 acres, and that of the semicircle is 39 916 800 square ft.; these two numbers are respectively the product of the numbers 1 to 6 (factorial 6, written 6!) and the product of the numbers 1 to 11 (factorial 11, written 11!).

The next stage in the construction of the New Jerusalem is to draw within its square the circle of diameter 7920 which represents the earth, average diameter 7920 miles. Tangent to this circle, with their centres on the larger circle at the four cardinal points, four smaller circles are drawn. The diameter of each is equal to the difference between the side of the New Jerusalem square, 7920, and the diameter, 10 080, of the circle which has a perimeter equal to that of the square. That difference is 2160, so the four small circles represent the moon which has a diameter of 2160 miles.

On either side of each of these four lunar circles, not quite touching them but with their circumferences passing through the eight points where the New Jerusalem square and circle intersect, eight similar circles are added, making twelve in all. They figure the twelve foundations, placed north, south, east and west in the New Jerusalem, and they are also its twelve jewels, as well as symbolizing the twelve moons (diameter 2160 miles) in the lunar year and the twelve great months, each of 2160 years, in which the sun runs its course through the zodiac. An outer circle of radius 6120 is then drawn to enclose the ring of twelve 'moons'.

This framework will accommodate many other schemes and orders of geometry and its further development is less firmly defined. It is apparent however from both the Stonehenge and Glastonbury plans that within the circle representing the earth, diameter 7920, should be drawn, by means of a hexagon, a circle of half that size, with a diameter of 3960. Within this circle the Glastonbury scheme (figure 4) shows another, of diameter 2160, which thus represents a thirteenth 'moon' round the centre of the diagram.

The construction of this or any other centralized figure creates a dot in the very middle of the design where the point of the compass has rested. It is good practice in geometry, where several lines have to be drawn through the centre of a figure, to withdraw the pencil just before the dot is reached, leaving it clearly defined and creating a small circular space round it. In

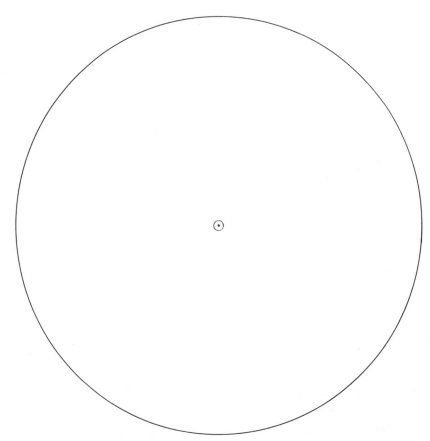

Figure 11. Function of the universal pole in measuring the areas of the New Jerusalem.
central dot, radius 12, area   = 1 unit
small circle, radius 72, area   = 36 units
earth circle, radius 3960, area = (if π = 22/7) 108 900 units
= (if π = 864/275) 108 864 units

The difference between the two figures for the area of the earth circle in the New Jerusalem is 36 units, equal to the area of the small circle round the universal pole.

sacred geometry either the central dot or the circle is taken to represent the pole of the universe, passing vertically through the centre of the diagram. The pole of the universe is the inner symbol of the canon, the fount of orthodoxy and the standard of universal law. It provides the yardstick by which the sacred diagram is measured, and since 12 is the number which

pole of universe
inner symbol of canon

measures the New Jerusalem, the radius of the central dot is taken as 12 units in length. Figure 11 illustrates the cross-section of the pole, its function in measuring the areas of the City and the reason why the radius of the circle surrounding the centre is taken to measure 72 ft.

Finally the diagram is enclosed within a twelve-sided figure, each of its sides touching one of the twelve lunar circles. Four of its sides, north, south, east and west, are those of a regular dodecagon, while the other eight are slightly shorter due to the positioning of the eight lunar circles they are required to touch. Calculating the dimensions of this figure is a nice problem for mathematicians. Both the linear dimensions and the areas of the various figures in the New Jerusalem are framed in units of twelve — 12 furlongs, 12 hides, 14 400 cubits, 1440 acres and so on — and the number twelve should therefore be exhibited in the dimensions of the twelve-sided figure. Suspecting that its area might be 120 million square units, but lacking the mathematical skill to prove it, the author sought the help of the late Ian Sommerville, who confirmed that that was in fact the case. The problem was inherited by Robert Forrest, who has made his name as a debunker of Velikovsky and other unorthodox theorists, but has the patience and good nature to help those among them who need it with mathematical advice. He refined Sommerville's figures and computed that, with the four regular sides of the dodecagon equal to 3280 and the eight others each of 3271.96 units, its exact area was 120 002 064 square units. Thus the entire structure of duodecimal number is contained in an area of virtually 120 000 000.

Figure 12 shows the construction of the New Jerusalem squared circle by the method described on page 70, involving a double Vesica. The perimeters of both the square and the circle contained by the Vesicas measure the canonical 31 680, and the diameter of the circle is 10 080. Multiplying that figure by the square root of 3 (taken as 97/56) gives the height or longer axis of the Vesicas. This proves to be 17 460, or ten times that most significant of symbolic numbers, the Number of Fusion, 1746. The gematria of this number (discussed more fully in a later chapter) confirms the identity of our New Jerusalem diagram, for $1746 = ’Ιερουσαλημ, ἡ πολις Θεου$, Jerusalem, the City of God.

As an ideal model of the cosmos the New Jerusalem is mathematically comprehensive, expressing within its simple geometric form the base numbers of duodecimal notation, the numbers from 1 to 12. The area of its circle, diameter 10 080, is twice 11!, or 79 833 600. Only the number 12! is

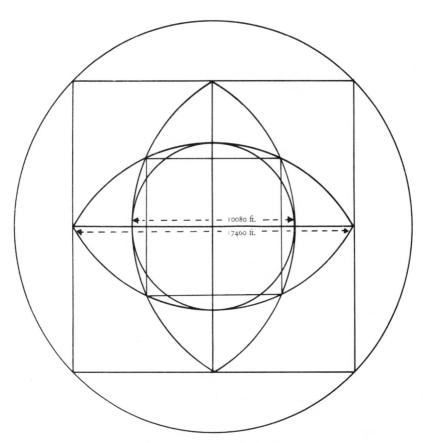

Figure 12. The squared circle of the New Jerusalem (radius 5040 ft.) contained by a double Vesica, a figure which has a length √3 times greater than its width. The length of the Vesica is therefore 17460 ft., ten times the 'number of fusion' (1746) which gives by gematria the phrase 'Jerusalem, the City of God'. Adjusted to 17461.96 ft., the length of the Vesica becomes the width of a square containing 7000 acres. The circle containing the square of 7000 acres has an area (if $\pi = 864/275$) of 12!, or $1 \times 2 \times 3 \times 4 \times 5 \times 6 \times 7 \times 8 \times 9 \times 10 \times 11 \times 12$ sq. ft. exactly. Thus:

$$\text{area of New Jerusalem circle} = 2 \times 11! \text{ sq. ft.}$$
$$\text{area of outer circle} = 12! \text{ sq. ft.}$$

missing; and it is discovered in the expanded figure. A square is drawn to contain the double Vesica, and the area of that square is 7000 acres. In that case, the circle which encloses the square (calculating with $\pi = 864/275$) has an area of 12! square feet exactly.

The numerical code of the New Jerusalem develops from the first eleven numbers through the number 720 or 6!, as below.

$$1 \times 2 \times 3 \times 4 \times 5 \times 6 = 720$$
$$8 \times 9 \times 10 = 720$$
$$720 \times 7 = 5040, \text{ the radius of the circle}$$
$$720 \times 11 = 7920, \text{ the side of the square}$$
$$5040 \times 7920 = 11!, \text{ half the area of the circle}$$

The series of the first twelve numbers is completed by the outer circle in figure 12, whose area is 12!. The diagram thus accommodates every order of number up to 12 and can be divided up into the greatest possible number of integral parts. In the same way it accommodates every order of geometry. This enquiry is limited to the plan of the New Jerusalem, but St John describes it as three-dimensional. Its solid form is a dodecahedron with twelve pentagonal sides, displaying the geometry of the number five. The plan is plainly adapted to receive the geometry of the numbers 3, 4, 6 and 8, and illustrated in figure 39 is the conformity of seven-fold geometry with the plan of the New Jerusalem.

Measuring the courts of the City

The linear dimensions of the New Jerusalem having been established, the next stage in its analysis is to measure the areas of its various parts or courts. This is made plain in Revelation 11, where St John is given 'a reed like unto a rod' and told to 'measure the temple of God, and the altar, and them that worship therein'. He is also told that from the total reckoning he should omit a certain area. 'But the court which is without the temple leave out, and measure it not . . .'

For measuring the courts of the New Jerusalem so that each of them is represented by a whole number of symbolic significance, the appropriate standard is found to consist of a circle with radius of 12 ft. A circle of that size is identified with the dot made by the compass point at the centre of the diagram. Its area refers once more to the number 3168 through being equal to a seventh part of 3168 sq. ft. This unit of 3168/7 sq. ft., equal to the area of a circle with radius of 12 ft., is here referred to as the New Jerusalem (NJ) unit.

The NJ unit is related geometrically to the acre in that a circle containing 3025 NJ units has a radius of 660 ft. or one furlong, and the square

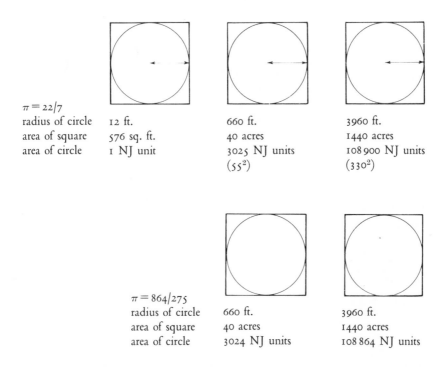

$\pi = 22/7$

radius of circle	12 ft.	660 ft.	3960 ft.
area of square	576 sq. ft.	40 acres	1440 acres
area of circle	1 NJ unit	3025 NJ units	108 900 NJ units
		(55^2)	(330^2)

$\pi = 864/275$

radius of circle	660 ft.	3960 ft.
area of square	40 acres	1440 acres
area of circle	3024 NJ units	108 864 NJ units

Figure 13. The New Jerusalem (NJ) unit in relation to the acre.

enclosing it has an area of 40 acres. The New Jerusalem square of 1440 acres contains a circle of area 108 900 NJ units, and a circle with circumference equal to the perimeter of that square has an area of 176 400 or 420^2 NJ units. In these calculations π is 22/7.

If, however, the closer value to π, 864/275, is used, the results are subtly different. The acre and the NJ unit are still related, but now the square of 40 acres contains a circle of 3024 NJ units, and the circle inscribed within the square of 1440 acres contains 108 864 rather than 108 900 NJ units. The difference of 36 units, slight though it is, proves to be of crucial significance when it comes to analysing Plato's numerical structures.

The difference between the two results, 36 NJ units, is the area of a circle with radius 72 ft. With that circle placed at the centre of the diagram the results are equalized, the area in both cases being 108 864 NJ units. It is obtained either by multiplying the square on the radius by $\pi = 864/275$, or by using $\pi = 22/7$ and deducting the area of 36 NJ units from the result.

The use of $\pi = 864/275$ in the measurement of this circle is indicated by its dimensions as adapted from St John's specifications: diameter 12 furlongs or 7920 ft., circumference 14400 cubits or 24883.2 ft. Between these two dimensions the ratio is 864/275. The reason why 108864 is the appropriate measure of the New Jerusalem earth circle is that it is divisible by 504. The other main divisions of the diagram can also be divided by that number, the importance of which is seen in chapter 3, on Plato's adoption of the New Jerusalem diagram as the plan of his ideal city which he called Magnesia.

Figure 14 shows the rings of the New Jerusalem and their dimensions. The most convenient way of calculating the area of an individual ring is to subtract the square on its inner radius from the square on its outer radius, multiplying the remainder by 22/7. Thus the areas of the separate rings are:

central dot, radius 12 ft., area	1 NJ unit		
inner circle, width 60 ft., area	35 NJ units		
first ring, width 1008 ft., area	8064 NJ units	$= 504 \times 16$	
second ring, width 2880 ft., area	100800 NJ units	$= 504 \times 200$	
third ring, width 1080 ft., area	67500 NJ units	}	$= 504 \times 300$
outer ring, width 1080 ft., area	83700 NJ units	}	
total, radius 6120 ft., area	260100 NJ units		

The total area, deducting the 8100 units contained by the first ring, radius 1080 ft., is $252000 = 5040 \times 50$ NJ units.

St John's New Jerusalem is an image of the sublunary world, all that lies within the influence of the moon. Its various parts represent:

> inner circle of radius 72, the pole of the earth and universe
> circle of radius 1080, the underworld
> circle of radius 3960, the earth
> outer rings of width 2160, the heavens below the moon

The heaven and earth symbolism of the two main parts of the New Jerusalem is in accordance with the corresponding Chinese diagram called the River Map. A commentary on it in the classic Chinese *Book of Diagrams* says:

> The aggregate number of the Superior Principle of nature is 216 . . . and the aggregate number of the Inferior Principle of nature is 144 . . . The two aggregate numbers of the Superior and the Inferior Principles of nature, added together, make 360, the number of days (generally reckoned) for a year.

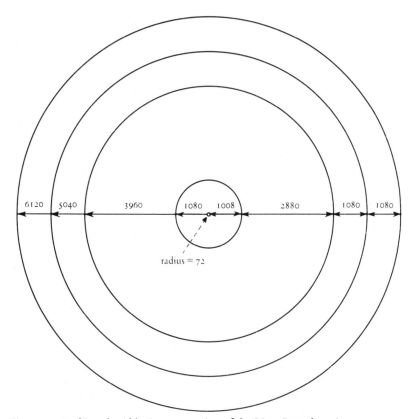

| 6120 | 5040 | 3960 | 1080 | 1008 | 2880 | 1080 | 1080 |

radius = 72

Figure 14. Radii and widths in cross-section of the New Jerusalem rings.

The ratio between the numbers 216 and 144 is 3:2 (a musical fifth), and this is also the ratio between the areas of two parts of the New Jerusalem, the ring of width 2880 ft. in cross-section and the outer pair of rings with combined width of 2160 ft. Their respective areas are 144 × 316800 sq. ft. and 216 × 316800 sq. ft. Numerically, therefore, these two parts of the diagram represent the Inferior Principle of earth and the Superior Principle of the heavens.

As an unlimited source of mathematical harmonies and felicities the New Jerusalem has obvious appeal to those who delight in such things. Yet its prestige in ancient times was far greater than would be attributed to mere curiosities of number. To the extent that its functions corresponded to those of the Chinese River Map, it provided a standard of proportion for the right conduct of human affairs and a means of attracting the benevolent powers of the cosmos. So it is stated by an ancient commentator on the *Book of Diagrams*:

> By means of the doctrine of numbers, virtuous conduct is brought into contact with invisible beings. The diagrams are useful in manifesting the right course of things to men and in bringing the virtuous conduct of men into contact with invisible beings. It follows, therefore, that when men harbour any doubts which they can not decide, the diagrams are then useful in the visible world, in the intercourse of men and in elucidating their doubts.

Similar claims are made by Plato on behalf of the number system which he weaves through the diagrams of his ideal cities and cosmologies: that its study refines and elevates the mind and thus leads to individual enlightenment and to the best possible forms of society. Plato's geometric and musical expressions of the cosmos, examined later, are all developments of the New Jerusalem scheme and its underlying pattern of number.

2 Number in sacred science

IN THE DIMENSIONS of the New Jerusalem are encoded the science and cosmology of the ancient world. The code is numerical because number is capable of expressing the underlying relationships between and within all different classes of phenomena. In order to appreciate the significance of the New Jerusalem and to profit from its study, it is necessary to understand the principles of the language in which it has been transmitted, the language of symbolic number. The sections of this chapter, together with the essays in our fourth chapter on certain individual symbolic numbers, are intended as an introduction to the subject.

According to Plato there are five mathematical sciences: arithmetic, by which the plan of creation is most nearly apprehended; plane geometry, or the study of numerical proportions; three-dimensional geometry, which explores the nature of solid bodies at rest; astronomy, which has to do with those same bodies in motion; and music, which illustrates the harmonious composition of the universe. The study of number itself takes precedence over the other sciences because it is at the root of them all. And at the root of number itself, providing its basic framework, is a certain series of 'nodal' or 'universal' numbers which comprised the ancient sacred canon.

Arithmetic, as it was studied in the ancient world, is hardly recognizable as a relation of the subject applied today to the torture of schoolchildren. 'Profane' number was so despised by the Greek philosophers that they barely mentioned it in their writings. It was not even thought worthy of the name arithmetic, but was called logistics. In the scholium on Plato's *Charmides* (quoted by L.C. Karpinski in his preliminary notes to the *Arithmetic* of Nicomachus of Gerasa) logistics is scornfully defined as 'the theory which deals with numerable objects and not with numbers; it does not, indeed, consider number in the proper sense of the term . . . It has for its aim that which is useful in the relations of life and in business, although it seems to pronounce upon sensible objects as though they were absolute.'

That loyal Platonist, Thomas Taylor, in his treatise on *Theoretic Arithmetic*, referred crushingly to logistics as 'practical arithmetic, which though eminently subservient to vulgar utility, and indispensably necessary in the shop and the counting house, yet is by no means calculated to purify, invigorate and enlighten the mind, to elevate it from a sensible to an intellectual life, and thus promote the most real and exalted good of man'.

Taylor was following Plato, who insisted upon number as the first study for the rulers of his ideal state, because it frees the mind from attending to actual objects and raises it to the realm of abstract principles and archetypes. This advantage is not derived from 'mere commercial calculations'. Arithmetic, said Plato, is to be studied 'for the sake of pure knowledge'.

In the writings of the old philosophers there is common agreement that the true purpose of number is for investigating the universe. Traditional cosmogonies tell how the Creator dreamt up a scheme of number, and from that original thought everything else proceeded. From the dance of the atoms to the revolutions of the planets every type of growth and motion is governed by the same set of laws, which are the laws of arithmetic. Number was therefore called 'the first paradigm', the primary reflection of reality. The orthodox view of the matter was stated by Nicomachus of Gerasa in the first or second century AD. Nature, he says,

has been determined and ordered in accordance with number, by the forethought and the mind of him that created all things; for the pattern was fixed, like a preliminary sketch, by the domination of number pre-existent in the mind of the world-creating God, number conceptual only and immaterial in every way, but at the same time the true and eternal essence, so that with reference to it, as to an artist's plan, should be created all these things: time, motions, the heavens, the stars and all kinds of revolutions. . . .

[Number] existed before everything else in the mind of the creating God, like some universal and exemplary plan, relying upon which as a design and archetypal example the Creator of the universe sets in order his material creations and makes them attend to their proper ends. (*Introduction to Arithmetic*, I, 6 & 4.)

In their cosmological inquiries the ancients paid less attention to physical phenomena than to the study of number. In the fixed ratios of arithmetic they discerned a constant reality ('that which always is and never becomes') as distinct from the apparent but illusory world of phenomena ('that which is always becoming but never is'). It is commonly supposed that their attention to ideal rather than to observational astronomy must have produced an inaccurate view of the world. Indeed, the records of the Greek astronomers prove that many of their speculations were considerably wide of the mark. They acknowledged, however, that their predecessors in past ages were better informed than themselves in all branches of science. That appears to have been the case, for the astronomical and other data, encoded in the traditional canon of number, indicate that the level of knowledge in remote antiquity was far higher than in early historical times.

The Pythagoreans, who were dedicated by their leader to the task of reassembling the scattered wisdom of the past, have passed down valuable information about the ancient approach to number. Each number symbolized one of the gods or dynamic tendencies in nature, and these were studied in relation to the characteristics of their numerical equivalents. The division of numbers into the odd and the even reflected the division of the universe into positively and negatively charged elements, odd numbers being positive or active and even numbers negative or passive. Their polarity is made explicitly sexual in Plutarch's *Roman Problems*, where odd numbers are said to be male because, if any odd number is set out as a row of units, the central point in that row is marked by a unit or part, whereas in a row of an even number of units the centre is a void.

Even numbers were further divided into the evenly-even (those in the series 2, 4, 8, 16, 32, 64 etc., the powers of 2), the evenly-odd (such as 6, 10, 14, 18, 22 which are composed of two odd-number halves) and the unevenly-even, which can be divided once or several times into even-number halves but eventually resolve themselves into odd numbers; examples are 24 and 28.

Many other categories of number are given by the old mathematical writers. Theon of Smyrna describes the properties of primes and composites, square, oblong, unequilateral, equally-equal, unequally-equal, planar, circular, parallelogrammatic, triangular and other numerical types. Their names commonly refer to the shapes formed by numbers when displayed as patterns of dots or pebbles. We still speak of square numbers, those produced by any number multiplied by itself, though we are no longer so inclined to visualize them literally as square in form. The old method of teaching arithmetic, by showing the relationship of numbers to shapes, was designed to illustrate basic principles and to lead the minds of children into the habits of reason. One can learn by rote that square numbers are formed by adding the next in the series of odd numbers to the previous square (e.g., $1^2 + 3 = 2^2$; $2^2 + 5 = 3^2$; $3^2 + 7 = 4^2$; $4^2 + 9 = 5^2$), but the formula only comes to life when one sees it in action, as in figure 15.

49

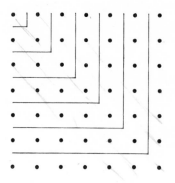

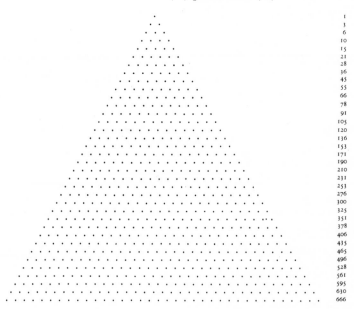

Figure 15. The square numbers, 1, 4, 9, 16, 25, 36. Sucessive odd numbers, 3, 5, 7, 9, 11 etc., constitute the differences between each square number and the next in the series.

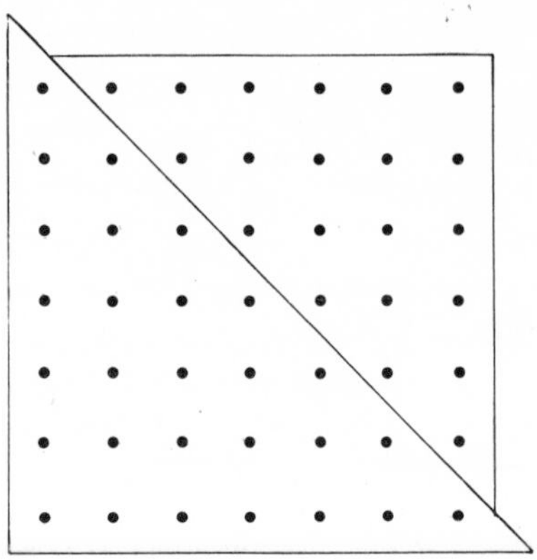

Figure 16. Every square number is the sum of two successive triangular num, bers, the greater of which is the number in the triangular series corresponding to the root of that square number. Thus 36 (6²) is made up of the sixth triangular number (21) plus the fifth (15).

1
3
6
10
15
21
28
36
45
55
66
78
91
105
120
136
153
171
190
210
231
253
276
300
325
351
378
406
435
465
496
528
561
595
630
666

Figure 17. The first 36 triangular numbers proceeding from the Tetractys (1 to 10) to the number 666.

Every square number is made up of two successive triangular numbers, and a triangular number consists of all the numbers from unity up to a given limit; the series they form is $3 (= 1+2)$, $6 (= 1+2+3)$, $10 (= 1+2+3+4)$ and so on. All such numbers can be displayed as an equilateral triangle made up of units. The Tetractys, the triangular form of ~Tetractys~ the number 10, had the same significance to the Pythagoreans as the Tree of Life diagram has in Jewish mysticism, both being accounted symbols of the universe, and the numbers 1 to 4 of which it is composed were said to be at the root of all creation. The apparently inconsequential words with which Plato begins the *Timaeus*, 'One, two, three, but where is the fourth?', have been claimed as an allusion to the primary importance of the Tetractys in the numerical philosophy of Pythagoras.

Numbers were also characterized by their parts or factors. This mode of analysis revealed 'perfect' numbers, those which are equal to the sum of their factors such as

$$6 (=1\times 6 = 2\times 3 = 1+2+3)$$
$$28 (=1\times 28 = 2\times 14 = 4\times 7 = 1+2+14+4+7).$$

Those which are exceeded by the sum of their factors, such as 12, were called 'superabundant' while those with factors amounting to less than themselves were said to be 'deficient'. Any two numbers of which the sums of their respective factors add up to each other were called 'amicable'. Pythagoras referred to the smallest pair of such numbers when he described friendship as similar to the relationship between 220 and 284.

The proliferation of numerical categories meant that each number had a variety of properties which, taken together, defined its natural character. Different systems of number symbolism were taught by the ancient and early Christian philosophers, but it was commonly held that since all other numbers proceed from the first ten, the decad contains the story of creation and how it developed.

The primary division of numbers is traditionally into two classes, the first of which comprises the monad, number one, and the second all other numbers. Being uniquely indivisible into integral parts, the monad is the image of that 'whole of wholes', Plato's name for the universe. Plotinus indeed made it even more transcendental as the symbol of the primal being, referred to by the Hebrew cabalists as *En Soph*, of whom nothing meaningful can be said and whose nature is beyond conjecture. The monad, according to Plotinus, contains in potential all numbers and gives them existence, but unless acted upon by its numerical emanations remains ever motionless and unapparent. For such reasons it was placed in a class by itself and was regarded rather as a symbol than an actual number. In the words of the old chant, thought by some to be of Druid origin, 'One is One and all alone and ever more shall be so.' The Greeks related the monad to Chaos, the state in which the universe existed before the appearance within it of two opposite tendencies, symbolized by the dyad, and the consequent beginning of motion and matter.

Two, the dyad, represents the lapse from unity, the split between heaven and earth, the rivalry and mutual dependence of the negative and the positive and all other pairs of corresponding opposites. The symbol of the first active stage in creation, its equivocal nature arises from the conflicting desires of its two parts, to react against each other and to seek reunion.

Three is the first odd, active or male generative number, and the first representative of fertility in nature. It is the number of the second stage in creation, the productive union between the negative and the positive which follows the separation and refinement of these opposite elements. The ancient veneration of the triad as the symbol of God in nature is renewed in the doctrine of the Trinity.

Four, the first square number, reconciles the two forms of mathematical growth, being both 2 + 2 and 2 × 2, and it represents the human instinct for symmetry and order by dividing the compass into four points and the year into four seasons. It is at the foundation of civilization, settlement and rectangular land division. It is the foursquare number of solid earth as opposed to the formless heavens. The tetrad is a symbol of completion within the decad, ten being the sum of the numbers one to four.

Five, the pentad, is called a marriage number because it is the sum of 2 and 3, the first even and the first odd numbers, and it constitutes a symbol of justice and moderation through providing the arithmetical mean within the decad. The correspondence of the five points of a pentagram to the five extremities of the body makes it an emblem of health and of humanity.

Six, the hexad, is the first evenly-odd number and it is also perfect, oblong and (because the powers of 6 always end with 6) circular. Nicomachus called it the marriage number because it is the first product of an odd and even number (3 × 2). The numbers 6 and 12 frame the proportions of the heavenly bodies, divide up the circle and measure the periods of solar time.

Seven is unique among the numbers of the decad because, as the Pythagoreans said, 'it neither generates nor is generated', meaning that it can not be multiplied to produce another number within the first ten, nor is it the product of other numbers. For that reason the heptad was called the Virgin and was a symbol of eternal rather than created things. It was particularly related to the measurement of time, the seven ages of man and the seven days of the week or quadrant of the lunar month. It is thus connected with the perfect number 28 which governs the periods of the moon and the female, and it is also the cause of 28 being a triangular number, for $1+2+3+4+5+6+7=28$.

Eight, the ogdoad, is the first cubic number ($2 \times 2 \times 2$) and stands for the cube, which has eight vertices and as the most stable of solid figures is the symbol of the element earth.

Nine, the ennead, is associated with completion and limit because it terminates the series which begins again with ten.

Ten itself, the decad, is a lower form of the monad, representing the created universe. As a number of this world with many practical qualities, it has attracted veneration in modern times. Its role in the rise and fall of Atlantis is reviewed in a later chapter.

Ancient theology was a science based on number, and its propositions were formulated through the medium called arithmology. The names of gods and abstract principles were linked to numbers within the decad which seemed to reflect their attributes. The monad was the obvious symbol for the solar principle in every system, as the sun in relation to the planets or Apollo among the Greek gods. To the dyad, symbol of duality and matter, was allotted a variety of gods, including Zeus, Demeter and Artemis. The triad was compared to the three-fold Hecate and to Leto the mother of Apollo (one) and Artemis (two); the tetrad to Hermes, Hephaestus, Hercules and Dionysus, the pentad to Aphrodite and Justice, and the perfect hexad to Virtue. Seven, the Virgin heptad, was appropriate to Athena who was not born from a womb but sprang from the head of Zeus in full armour and brandishing weapons. The cubic ogdoad, number eight, was attached to Rhea/Cybele, and was taken by the gnostics as a symbol of Jesus. The ennead, number nine, was associated with the names of Hera, Prometheus, Ares and many others, while the decad embraced all ten numbers under the name Cosmos.

The numerous correspondences discovered between the various gods and the first ten numbers, of which the above is a mere sample, seem in many cases obscure and arbitrary, though the reasons behind them were no doubt clear enough to the professors of arithmology. This, however, was but the

53

start of the matter. For a more accurate representation of the peculiar attributes of each god larger numbers were required. They were generated from harmonies and ratios within the decad, in combination with other numbers beyond it, to produce a universal scheme or numerical pantheon in which every deity had its place according to its characteristic number. That scheme is the subject of this present inquiry, beginning here with its bare bones, the skeleton of number that underlies all the products of ancient science and philosophy, and proceeding to analysis of its parts, its applications and its overall meaning.

The numbers of the canon

In the operations of simple arithmetic and throughout all the numerical manifestations of nature, such as the periods and intervals of the solar system, certain 'nodal' numbers occur, providing a link between processes and phenomena which otherwise appear quite unconnected with each other. Most prominent among these are the multiples of 72, including the powers of 12 and numbers such as 5040 which is the product of the first 7 numbers multiplied together, written as 7!.

$720 = 72 \times 10 = 6!$	$3456 = 72 \times 48$
$864 = 72 \times 12$	$3600 = 72 \times 50$
$1008 = 72 \times 14$	$3960 = 72 \times 55$
$1080 = 72 \times 15$	$4320 = 72 \times 60$
$1152 = 72 \times 16$	$5040 = 72 \times 70 = 7!$
$1224 = 72 \times 17$	$5184 = 72^2$
$1296 = 72 \times 18$	$5760 = 72 \times 80$
$1440 = 72 \times 20 = 12^2 \times 10$	$6120 = 72 \times 85$
$1512 = 72 \times 21$	$6336 = 72 \times 88$
$1584 = 72 \times 22$	$7920 = 72 \times 110$
$1728 = 72 \times 24 = 12^3$	$10368 = 72 \times 144$
$2160 = 72 \times 30$	$20736 = 72 \times 288 = 12^4$
$2520 = 72 \times 35$	$40320 = 72 \times 560 = 8!$
$2592 = 72 \times 36$	$108864 = 72^2 \times 21$
$2664 = 72 \times 37$	$114048 = 72^2 \times 22$
$2880 = 72 \times 40$	$248832 = 72^2 \times 48 = 12^5$
$3024 = 72 \times 42$	$362880 = 72 \times 5040 = 9!$
$3168 = 72 \times 44$	$39916800 = 72^2 \times 7700 = 11!$

In these numbers are the dimensions and distances of the earth, sun and moon, the measurements of the ideal universe and the formula behind many ancient expressions of it, such as: the New Jerusalem diagram, the plan of the Stonehenge temple, Plato's perfect city in the *Laws*, the scale of music by which he represented the universal soul and the order he discerned among the planets. Through the mystical science of gematria (described in the following chapter) these same numbers provided some of the principal sacred names of Christianity, including the title of its founder, Lord Jesus Christ, which has the value by gematria of 3168, the key number in the New Jerusalem dimensions.

3168

Multiples of 72 are not the only kind of number with traditional symbolism; others such as 666 and 1746 are reviewed in a later section. But those listed above are predominant in traditional cosmologies and religious systems, not as a matter of convention but because of their natural prominence in the operations of arithmetic. Their marvellous properties become apparent as they are investigated and it is seen how they act as nodes, marking the intersections of different numerical orders and structuring the whole field of number itself.

666

1746

The geometric symbol which generates and contains all the nodal numbers of the Canon has at its centre the basic figure of the New Jerusalem, a circle of radius 5040 and circumference 31680.

$r = 5040$

$c = 31680$

The use of number to reconcile apparently disparate types of phenomena is mentioned several times by Plato, but always in the obscure manner in which he treated the most important, esoteric part of his subject. One such reference is in the *Epinomis*, a kind of appendix to the *Laws*.

The most important and first study is of numbers in themselves: not of those which are corporeal, but of the whole origin of the odd and the even and the greatness of their influence on the nature of reality. When these things have been learnt, there comes next what they call by the ridiculous name of 'geometry' [literally 'land-measuring'], when it proves to be a manifest likening of numbers not like one another by nature, in respect of the province of planes; and this will be clearly seen by him who is able to understand it to be a marvel, not of human but of divine origin. And then, after that, the numbers thrice increased and like to the solid nature, and those again which have been made unlike he likens by another art, namely that which its adepts call stereometry . . .

Every diagram and system of number and every combination of harmony and the agreement of the revolution of the stars must be made manifest as one in all to him who learns in the proper way, and will be made manifest if a person learns aright by keeping his eyes on unity; for it will be manifest to us as we reflect, that there is one natural bond linking all things.

The main components of that code of number, the universal bond on which Plato based his philosophy, are listed above, and those which are most closely linked with his city of Magnesia and the New Jerusalem are reviewed later. Their study, so Plato claimed, is uniquely efficacious in refining the mind and attuning it to proportion and justice. He saw it also as a means to rediscovering that long-neglected canon of harmonies which illuminates the human soul and provides a true standard for public affairs.

Gematria: the names and numbers of God

In the heroic days before settlement and the building of temples and cities, the tribes of Israel travelled between the sacred places of their territory, carrying with them the Ark of the Covenant and the materials of its resting-place, the Tabernacle. At the traditional spots where they pitched camp, they erected the Tabernacle, a series of curtained enclosures with the Ark lodged in its inner sanctuary, and set up their tents in ritual order around it. The Temple at Jerusalem was designed in imitation of the Tabernacle.

Religion in those days was an entirely practical affair. The wandering tribes depended for their livelihood on communion with the native spirits of their landscape, and the Tabernacle was designed, like the Temple after it, to attract by similitude the powers of divinity. The names given to the various deities, or the various aspects of the vital principle in nature, were based on the sounds which were found most effective in evoking their response; from these sounds arose the alphabet. The legendary architect of the Tabernacle, Betzal'el, was said to have known 'the combination of letters with which heaven and earth were made', and to have fashioned the Tabernacle in accordance with them.

Each letter represented a particular type of universal energy. It also corresponded to the shape, colour, perfume and other characteristic attributes of the god it symbolized, and most essentially to the god's number. When temples were built as instruments of invocation, they expressed in their dimensions, in their furnishings and in the areas and shapes of the spaces they enclosed, the number of the god to whom they were dedicated. Thus the temple of Athena the Virgin was laid out in units of the virgin seven. Betzal'el, and architects in the masonic tradition who followed him, were initiated into the science of acoustics, and the ultimate product of their craft, the cosmic temple, was designed to resonate with the music of planetary motion. In its proportions were to be found every type of musical harmony, set out in numerical ratios; and since these numbers corresponded to sounds and letters, the whole structure was an architectural litany, containing the names by which all the various powers in nature were invoked, and thus forming a pantheon, a complete representation of Universal Mind.

Plato's *Cratylus*, the most deeply mystical, least understood of all his works, consists of an inquiry into the origin and meaning of names. The learned Cratylus claims to have some special knowledge which tells him that the names of objects are of divine origin and are not merely formed by human convention. A sceptic, Hermogenes, debates the matter with Socrates and objects to the theory on the grounds that things are called by different names in different languages, so it is impossible to say that any particular name is 'correct'. Socrates points out that certain sounds do appear to have a consistent meaning. The letter 'r' for example conveys a sense of disruption and rush, as in the words 'run', 'tremble', 'break', 'rend', 'crush', and 'whirl', while 'l' is liquid and flowing, 'g' is gummy and glutinous and 'n' is associated with inwardness. This observation is made sharper today by the fact that most of the Greek words with which Socrates illustrates the idea retain the same characteristic letter when translated into English. He admits that words are often quite different in other languages, but says that this does not rule out the theory that sounds possess archetypal meanings. The wise men of every nation, who first gave names to things, might have derived them all from the same natural source, but with different results. In the same way, artists who draw from the same model produce pictures which are entirely different from each other. In the course of his argument Socrates makes an odd statement:

> So perhaps the man who knows about names considers their value and is not confused if some letter is added, transposed, or subtracted, or even if the force of the name is expressed in entirely different letters.

In this and other passages of the dialogue there is a suggestion that Socrates or Plato had in mind that strange old cabalistic science which inquires into the hidden meaning of names, particularly those of gods and sacred principles, by reference to the numerical value of the letters comprising them. Many languages retain from very early times an alphabet in which every letter is a symbol, a sound and a number. The Jews, who have cherished their old culture and cabalistic lore, are inclined to pretend that their tradition is the oldest of all – as they are entitled to do and as others may equally dispute. For there is no knowing where, or among which people, the institution of the Temple and the science of computing the names of God first made their appearance. The Greeks honestly confessed their ignorance of the matter, and Plato accepted the conventional view of his time, that culture and science were introduced by a race of divine beings who once ruled on earth.

The most topical cause for interest in the cabalistic practice, known as gematria, of interpreting the names and attributes of God by means of letter and number, lies in the use made of it by the founders of Christianity. That they did so use it, and that many of the epithets of Christ and other elements in the Christian legend correspond numerically to the sacred names woven into the plan of the cosmic temple, can be established beyond reasonable doubt by applying gematria to the original Greek of the New Testament. The fact is confirmed by the best of authorities, certain early Christian fathers, who knew only too well that many of the principal names in their faith were based on a code of number, because they waxed furious in writing against those of their co-religionists who taught or studied the subject or attached any significance to it.

Their chief opponents were the gnostic masters, such as Marcus and Valentinus of the second century AD, who sought to preserve the pristine spirit of Christian enlightenment during the period of its decline. The gnostics were so called because they claimed to have the gnosis or knowledge through personal experience, acquired by initiation into a reformed version of the pagan mysteries. Gnostic religion developed at the start of the Christian era, probably in Alexandria, from a fusion of Hebrew mysticism, Greek philosophy and the hermetic traditions of Egypt and the East. By the second century it was under pressure from the authorities of the Christian Church, now established in the capital of its old enemy, imperial Rome, and the gnostic sects were persecuted into oblivion.

One of the points of difference between the gnostics and the Church was whether or not Christianity should acknowledge any debt to previous forms of religion. The gnostics were proud of the knowledge they had inherited from the ancient world, and claimed that the numerical science of the pagan philosophers could be adapted to prove the truth of Christianity. The Church, on the other hand, taught that the coming of Christ was a unique event which had raised human understanding above its previous level, rendering all earlier religions obsolete. Roman policy was to extirpate the traditions and records of pagan science, particularly among Christians, with the result that information about the gnostics' number theology is nowhere to be found but in the works of those Church Fathers, Irenaeus, Tertullian and Hippolytus above all, who wrote in order to discredit it.

The most informative of them is St Irenaeus, the author of a five-volume work *Against Heresies*. He was Bishop of Lyons in the second century, and his speciality was exposing the evils of gnosticism. His manner of writing was like that of a scandal-monger journalist who presents a selective version of his victims' point of view in order to make them look absurd or sinister. Against the gnostics he brought up all the accusations which are typically applied to unorthodox cults and their leaders: that they delude and swindle their pupils, disrupt family life, seduce their women followers and presume to teach other doctrines than those officially prescribed. It may be that Irenaeus was instructed for a time by the gnostics and then rejected from one of their schools, for he claims it to his credit that the gnostics judged him ignorant. Since his object in writing was to ridicule the gnostics' science, he could not be expected to provide a sympathetic account of it, and he gives no indication of the philosophy behind their use of numbers for scriptural exegesis. Yet his evidence is enough to justify the contention that certain books and passages in the New Testament, products of early gnosticism, have an inner meaning which can only be elicited through numerical analysis, the method which the cabalists call gematria.

The numerical equivalents of the Greek letters, as given in any lexicon, are as follows:

$A \, \alpha$	$B \, \beta$	$\Gamma \, \gamma$	$\Delta \, \delta$	$E \, \varepsilon$	$Z \, \zeta$	$H \, \eta$	$\Theta \, \theta$
1	2	3	4	5	7	8	9
$I \, \iota$	$K \, \varkappa$	$\Lambda \, \lambda$	$M \, \mu$	$N \, \nu$	$\Xi \, \xi$	$O \, o$	$\Pi \, \pi$
10	20	30	40	50	60	70	80
$P \, \varrho$	$\Sigma \, \sigma, \varsigma$	$T \, \tau$	$Y \, \upsilon$	$\Phi \, \varphi$	$X \, \chi$	$\Psi \, \psi$	$\Omega \, \omega$
100	200	300	400	500	600	700	800

There were formerly two other letters, for 90 and 900, but they became obsolete in literature and were retained only as numbers. The digamma, standing for the number 6, fell entirely out of use and was replaced for numerical purposes by the symbol ς', which may also be used in place of the letters $\sigma\tau$ when they occur together. As examples, the number of the Beast in

Revelation (13,18) is written $\chi\xi\varsigma' = 600 + 60 + 6 = 666$, and $\sigma\tau\alpha\upsilon\rho\sigma\varsigma$, a cross, is either

$$200 + 300 + 1 + 400 + 100 + 70 + 200 = 1271, \text{ or}$$
$$6 + 1 + 400 + 100 + 70 + 200 = 777.$$

By the conventions of gematria one unit, known in Hebrew as *colel*, may be added to or subtracted from the value of a word without affecting its symbolic meaning. Thus $\dot{\epsilon}\kappa\kappa\lambda\eta\sigma\iota\alpha\ \theta\epsilon\sigma\upsilon$, 778, Church of God, is equivalent by gematria to $\sigma\tau\alpha\upsilon\rho\sigma\varsigma$, 777.

Words and phrases with the same number do not necessarily have identical meanings. Two words with phonetic similarities may have certain associations in common but each has its own function in language, and in the same way the names of gods and sacred principles retain individual significance even though they share the same number. They may indeed seem as diverse from each other as male from female or good from evil. But if their number is the same, they both refer to a particular power or active principle in the universe which the ancients symbolized by that number. John the Baptist, for example, who is $'I\omega\alpha\nu\nu\eta\varsigma\ B\alpha\pi\tau\iota\sigma\tau\eta\varsigma$, 2220, is identified, as the forerunner of Christ, with the spirit of prophecy, $\tau\sigma$ $\pi\nu\epsilon\upsilon\mu\alpha\ \pi\rho\sigma\phi\eta\tau\epsilon\iota\alpha\varsigma$, 2220, and the Christ-bearer, $X\rho\iota\sigma\tau\sigma\phi\sigma\rho\sigma\varsigma$, 2220. That number also belongs to the words with which St John's spirit-guide in Revelation 1, 11, announced its name: 'I am Alpha and Omega', $\dot{\epsilon}\gamma\omega$ $'A\lambda\phi\alpha\ \kappa\alpha\iota\ '\Omega\mu\epsilon\gamma\alpha$. An esoteric Christian symbol of the principle behind the number 2220 is the grain of wheat (John 12, 24), $\dot{\sigma}\ \kappa\sigma\kappa\kappa\sigma\varsigma\ \tau\sigma\upsilon\ \sigma\iota\tau\sigma\upsilon$, 2220, which must fall into the ground and die, after which 'it bringeth forth much fruit'. All these terms have different associations, but they have a common reference, being the various symbols of that eternal prophetic spirit which lies dormant for long ages, but periodically revives its energies to refresh human spirits and culture.

Where we can be sure that gematria was involved, as in the case of the holy names of Christianity, the likelihood of chance coincidence of numbers is diminished. Less certain is the extent to which numbers were formerly applied to more general terms in secular use. Some would say that words tend to be formed from the appropriate sounds, as Socrates suggested in *Cratylus*, and thus spontaneously acquire the appropriate number. On the other hand, a word such as Pneuma, $\pi\nu\epsilon\upsilon\mu\alpha$, meaning breath, which shares its number, 576, with $\dot{\alpha}\epsilon\tau\sigma\varsigma$, eagle, may have been adopted as the religious term for Spirit for mathematical reasons, because $576 = 24^2$.

The number of the name Jesus, 'Ιησους, is 888. According to Irenaeus the gnostics therefore referred to him as the Ogdoad or number 8. Their Pleroma, or hierarchy of divine powers, consisted of thirty aeons, divided into three groups of 8, 10 and 12. The first of these, the Ogdoad, represented the primal creation. Having six letters, the name 'Ιησους was also linked with the number 6, and the number 888 may be reduced to 6, because $8 + 8 + 8 = 24; 2 + 4 = 6$. Also, the number 888 was taken to correspond to the 24 letters of the Greek alphabet, of which there are 8 units, 8 tens and 8 hundreds.

888

Again illustrating the gnostics' use of number, Irenaeus wrote:

> The local positions of the three hundred and sixty-five heavens they distribute in the same way as the mathematicians, for they have taken their theorems and applied them to their own kind of learning. And their head, they say, is Abraxas, therefore he has in himself the three hundred and sixty-five numbers.

365

The significance of this is that the name Abraxas, 'Αβραξας, has the value by gematria of 365, and Abraxas was god of the 365 days of the solar year, corresponding to Mithras, Μειθρας, whose number is also 365, which is the number of 'ονομα άγιον, sacred name.

The prime number 37 was clearly of interest to the first Christians. Its most obvious numerological feature is that it generates the numbers in the series 111, 222, 333 etc., including the number of the Beast, 666 or 37 × 18, and 888 the number of Jesus, who is also the Founder ὁ οἰκιστης, 888, and who further indentified himself with that number by his claim, I am the Life, εἰμι ἡ ζωη, which has the value 888 or 37 × 24. In their pioneer work on gematria early this century, F. Bligh Bond and T.S. Lea showed that many of the sacred names and phrases in New Testament and gnostic writings have numbers which are multiples of 37. Subsequent research has turned up many further examples. Christ, Χριστος, has the number 1480 or 37 × 40, and

37

37 × 18 = 666

37 × 24 = 888

1480

1480 = Son of the Cosmos, υἱος κοσμου
holiness, ἡ ἁγιωσυνη
throne of wisdom, θρονος σοφιας
the twelve pearls, οἱ δωδεκα μαργαριται
Master and Lord (John 13,13), ὁ διδασκαλος, ὁ κυριος
the Perfector (an epithet of Christ), τελεσφορος
the hope of the kingdom of Jesus, ἡ ἐλπις βασιλειας 'Ιησου

The number of Jesus Christ, 888 + 1480, is equal to 37 × 64 or 2368, and that number also pertains to $\delta\iota\delta\alpha\iota\sigma\sigma\upsilon\nu\eta$ $\pi\iota\sigma\tau\epsilon\omega s$, righteousness of faith (Romans 4,13) and to $\dot{o}$ $\theta\epsilon os$ $\tau\omega\nu$ $\theta\epsilon\omega\nu$, the God of gods. In St John's Gospel (15,1) Jesus announces his identity through gematria, saying, 'I am the true vine, and my Father is the husbandman.' The true vine, $\dot{\eta}$ $\dot{\alpha}\mu\pi\epsilon\lambda os$ $\dot{\eta}$ $\dot{\alpha}\lambda\eta\vartheta\iota\nu\eta$, 558, added to Father the husbandman, $\dot{o}$ $\pi\alpha\tau\eta\rho$ $\dot{o}$ $\gamma\epsilon\omega\rho\gamma os$, 1810, makes 2368, Jesus Christ.

Among the Greek gods, the virgin Athena, $\dot{\eta}$ $'A\vartheta\eta\nu\alpha$, has the appropriate number 77, Hermes, $'E\rho\mu\eta s$, is 353 and Zeus, $Z\epsilon\upsilon s$, is 612, relating the head of the Greek pantheon to the Good Shepherd, $\dot{o}$ $\pi o\iota\mu\eta\nu$ $\dot{\alpha}\gamma\alpha\vartheta os$, 612. The ratios between the numbers of the gods allow their names to appear together in one geometric figure, as for example in the Vesica Piscis (figure 22), the basic figure of symbolic geometry. A Vesica with its width or shorter axis measuring 353, the number of Hermes, has a height or longer axis of 612, Zeus. If this Vesica is enclosed in another, the width of the larger figure is 612 and its height 1060, which is the number of $\pi\nu\epsilon\upsilon\mu\alpha$ $\vartheta\epsilon o\upsilon$, spirit of God, $\Sigma\iota\omega\nu$, Sion, and $\dot{o}$ $\dot{\alpha}\pi$ $\dot{\alpha}\rho\chi\eta s$, the God that existed from the beginning (I John 2,14). The addition of *colel* to 1060 makes 1061 which is the number of Apollo, $'A\pi o\lambda\lambda\omega\nu$, and several of his traditional epithets, such τo $\kappa\rho\alpha\tau os$, Might, and $\dot{o}$ $\phi\upsilon\lambda\alpha\xi$, the Guardian (given by Proclus). Terms with the number 1061 which were applied to Christ include $\upsilon\dot{\iota}os$ $\epsilon\dot{\iota}\rho\eta\nu\eta s$, son of peace (Luke 10,6) and the gnostic $\pi\eta\gamma\eta$ $\vartheta\epsilon o\tau\eta\tau os$, source of divinity, a reference to the Father in Christ.

The first principle of Christianity, Lord Jesus Christ, $K\upsilon\rho\iota os$ $'I\eta\sigma o\upsilon s$ $X\rho\iota\sigma\tau os$, has the number 3168, the most notable of all symbolic numbers, and the Holy Spirit, τo $\dot{\alpha}\gamma\iota o\nu$ $\pi\nu\epsilon\upsilon\mu\alpha$, 1080, took on both the number and attributes of the Earth Spirit, τo $\gamma\alpha\iota o\nu$ $\pi\nu\epsilon\upsilon\mu\alpha$, of which it is an anagram.

Irenaeus reported that the gnostics considered Jesus to have given proof of his divinity when he spoke the words, 'I am Alpha and Omega', because the sum of the letters alpha, 1, and omega, 800, is 801, and that is the number of a dove, $\pi\epsilon\rho\iota\sigma\tau\epsilon\rho\alpha$. This supported their belief that the body of Jesus was that of an ordinary man, into which the divine spirit, depicted symbolically as a dove, had descended at his baptism. Arguing against the teaching of the Church that the body and spirit of Jesus were equally divine, and that the faithful could look forward to bodily resurrection, the gnostics pointed out that, since divinity is beyond suffering, the agony of Jesus on the cross meant that his body was mortal. Beliefs such as this were the real cause of the persecution of gnosticism by Irenaeus and his colleagues. The Church had adopted as its totem the image of Christ's body, the god nailed to a cross. It was disliked by the gnostics for reasons similar to those given by the Protestants at the Reformation and by sects of spiritualist and charismatic Christians today: that it emphasizes the body and the material power of the Church rather than the spirit of Christ – a function somewhat like that of Lenin's mummified body in the state temple of his cult.

The religious symbols of the gnostics, as illustrated for example in Richard Payne Knight's *Worship of Priapus*, were of an entirely different nature. Like temple carvings in the East, their theme was sexuality, graphically and often humorously expressed. Gnostic artists exercised their inventiveness with the strange types of copulation they engraved on sigils and talismans, echoing the pagan custom of displaying images of sexual parts at religious festivals. Their purpose was to celebrate the sacred marriage between the two forms of vital energy in the universe, negative and positive or male and female; but the Church was out of sympathy with such traditional manifestations of religious feeling. They were considered to be regressive and a hindrance to the establishment of the new Christian code of morality, promoted by Rome under the sign of the Crucified God. In their rude parodies of this deified symbol the gnostic artists were sometimes outrageous, but the most telling stricture on it was delivered subtly by St John, using the numerical language of gnosticism, in Chapter 13 of Revelation. We return to this subject later, in an essay on the number 666 (p.185).

From Irenaeus's description of it, the gnostic practice of exegesis by number sounds, as he intended, vain and futile, and so it has been judged by theologians ever since. Yet several of the gnostic masters who taught it had great reputations for scholarship, and they were obviously sincere in believing that the numerical philosophy of the ancient sages was the greatest benefit that Christianity had inherited from the past. Far from controverting the truth of the new religion, it proved its legitimacy.

So said the gnostics, and now that they have passed away and gnosticism has long ceased to threaten the stability of organized religion, one is free to investigate their claims without prejudice or fear of rancour. Modern theologians agree that much of the content of the Gospels was adopted from earlier, pre-Christian religions, and Christianity is unshaken by such discoveries. Few would now think it discreditable that certain names, phrases and legends in the New Testament were based on that sacred code of number which informed the great religions and philosophies of antiquity. Indeed, after viewing the harmonious structure of ratios which was the prototype of St John's Holy City, one may come to sympathize with the understanding of the old gnostics, that Christianity is the richer, and the more worthy and likely to endure, for being rooted in the traditions of ancient sacred science.

Symbolic or sacred geometry

In the hierarchy of arts geometry ranks equal to music, expressing through visible shapes the same numerical proportions as musical harmonies present to the ear. Alberti, the fifteenth-century Renaissance architect, justified the use of musical themes in building by the Pythagorean doctrine that 'the numbers by means of which the agreement of sounds affects our ears with delight, are the very same which please our eyes and our minds'. Palladio related 'the proportions of measurements as harmonies for the eyes' to 'the proportions of voices as harmonies for the ears'. Such harmonies, he wrote, 'usually please very much, without anyone knowing why, except the student of the causality of things'. Thus, together with purely geometric ratios such as the side of a square to its diagonal ($1 : \sqrt{2}$) and the axes of the Vesica Piscis ($1 : \sqrt{3}$), initiated architects used the canonical intervals of music, the fourth ($3 : 4$), fifth ($2 : 3$) and octave ($1 : 2$), to proportion rooms, areas and entire buildings. These proportions were not derived directly from music, but from a common source in number as a symbol of the divine Creation.

Measurement of land is the literal meaning of geometry, and another of its practical uses, much emphasized in early military manuals, was for designing fortifications. Here the practical function of geometry merges with the sacred and magical. Mircea Eliade (*The Sacred and the Profane*) indicates that

> the fortifications of inhabited places and cities began by being magical defenses; for fortifications – trenches, labyrinths, ramparts, etc. – were designed rather to repel invasion by demons and the souls of the dead than attacks by human beings . . . . In the last analysis the result of attacks, whether demonic or military, is always the same – ruin, disintegration, death.

An example is the Great Wall of China, built by the rules of geomancy to protect the kingdom as much from evil demons from the north as from human invaders. Similarly, the extensive Iron Age earthworks of southern

England, enclosing or sheltering ancient sanctuaries, are often found to be impractical for defensive purposes and are presumed to be magical and symbolic. They correspond on a large scale to the geometric figures which magicians construct to protect themselves from their demons: and these same figures are discovered in the designs of ancient temples and places of invocation, beginning with megalithic circles.

Sacred geometry is the essence of the geometer's art. Its use is for making descriptions of the universe by combining together in one geometric composition the basic figures which represent the different orders of number and underlie the manifestations of nature. It is the art of synthesizing diverse elements. The tradition survives today in the patterns of Islamic art, but it is no longer thought worthy of scientific study on the grounds that it is irrational. That epithet is applied to mathematical ratios which can not be defined in terms of whole-number integers. Ever since the legendary renegade Pythagorean let the cat out of the bag by revealing that π, like many of the other important ratios in geometry, is irrational – and that one can not therefore define the relationship between the diameter and circumference of a circle or between the side and diagonal of a square by any pair of integers – the procedures of the old geometers, in squaring the circle and uniting different orders of geometry in one scheme, have been considered a mare's nest. Rationally indeed, the problem of constructing a circle of the same perimeter or area as that of a given square, or of fitting together geometric shapes developed from the numbers 5, 6 and 7, is insoluble because these shapes are not commensurate.

Yet there is one system in which all numbers and proportions and every disparate or incommensurable element co-exist and function together in perfect harmony, and that system is the universe. It can never be represented perfectly and literally by any artist's or geometer's model because, although it is said ultimately to be a creation of reason, the paradox in its nature is too deep rooted for human resolution. The geometer's aim, therefore, is to imitate the universe symbolically, depicting its central paradox by bringing together shapes of different geometric orders, uniting them as simply and accurately as possible and thus creating a cosmic image which, as Plato claimed for his Magnesia plan, 'most nearly resembles the original' and is 'second only to the ideal'. And within that limit of ambition it will be found possible, to all practical intents and purposes, to square the circle, unite the different orders of geometry and depict adequately that most perfect expression of the Monad which the Greeks called Cosmos.

The first figure of sacred geometry is the circle, whose circumference has neither beginning nor end and is therefore the geometer's image of entirety and eternity. As the simplest and most self-sufficient of space-enclosing shapes, and the matrix of all others, it is the natural symbol of that unique living creature, the cosmos. Thus Plato described it. The circle contains the polygons of every order of geometry, and it encloses a larger area in relation to the length of its perimeter than any other shape.

The ratio between the diameter and circumference of a circle is $1 : \pi$ or $1 : 3.141\,592\,654\ldots$, and the area of a circle is obtained by multiplying the square on its radius by π or by multiplying the radius by half the circumference. Since π is irrational, the relationship between a circle's width and its circumference or area can not precisely be expressed in whole numbers, so in the interests of rational cosmology it is given an approximate value as a fraction. Among the fractional values for π which appear to have been known in the ancient schools of geometry were:

the most convenient, $22/7 = 3.142857$
the most accurate, $355/113 = 3.141593$
the New Jerusalem π, $864/275 = 3.14\overline{18}$
also $3893/1080 = 3.14\overline{16}$
and $3927/1250 = 3.1416$

The symbolic opposite of the circle is the square. Whereas the circle represents the unknowable, spirit and the heavens, the square is material and of the earth. The ratio between its width and its perimeter, instead of being the irrational π as in the case of the circle, is simply and rationally 4. The circle and the square, made commensurable with equal perimeters, form a diagram of the fusion of matter and spirit and together illustrate human nature and the nature of the universe. The object of sacred geometry being to depict that fusion of opposites, the squared circle is therefore its first symbol. Temples and cosmological cities throughout antiquity were founded on its proportions.

'The belief that heaven is round and the earth square seems to have been held in China at an early date', says Lord Raglan the anthropologist in *The Temple and the House* (the chapter, 'Squaring the Circle'). He continues, quoting from J. Ross's *The Original Religion of China*, 'The altar to Heaven was enclosed in a circular space and that to Earth in a square.' The foundation legends of Rome describe the city as both a circle and a square. In Plutarch's *Romulus* its circumference is said to have been drawn as a ploughed furrow, the centre point being a circular pit called *mundus*, an apparent symbol of the earth with the pole of the universe passing through it. The word *urbs*, says Raglan, comes from *orbis*, round, suggesting that the city was originally circular. Yet in the other tradition of Romulus the city was Roma Quadrata and the primeval furrow was in quadrangular form.

It has been much debated whether the circle or the square came first in ancient sacred architecture. W.R. Lethaby wavered between the two and finally, in *Architecture, Nature and Magic*, gave priority to the circle. This seems an obvious conclusion. The circle has many forms in nature, from the orb of the full moon to a bird's nest, from the orbiting of the planets to the annual rounds of human and animal migrants. It is a symbol of orderly, repetitive motion, appropriate to times before settlement and reflected in the circular camps of nomads and the round forms of primitive shrines, huts and temples. The square, on the other hand, though it does have certain correspondences in nature, implies human construction and an attempt at permanence. Its adoption was evidently coeval with civilization and settlement, which it symbolizes. A city which is built simply as a square is a product of rational minds and lacks that element of imagination and divinity, symbolized by the circle, which is a necessary part of human life. Thus in every scheme of cosmological geometry the square and the circle are brought together and made commensurate.

The diagrams in figure 20 show the four ways in which relationships between the square and the circle can be depicted. In the first, the circle contained in the square has a diameter equal to the side of that square. In the second, the circle containing the square has a diameter equal to the square's diagonal, which is calculated by multiplying the side by the square root of 2 or 1.414 213 56 . . ., another irrational number, but almost perfectly defined by 19601/13860 (the mean between 140/99 and 99/70). The third

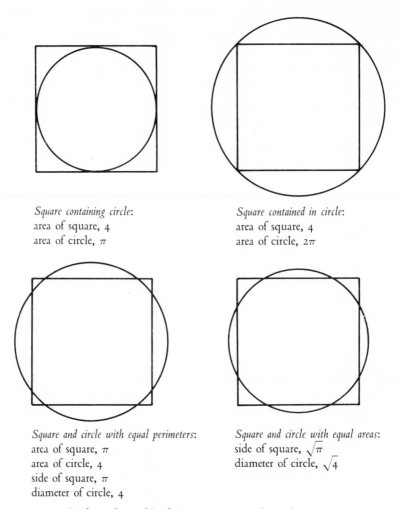

Square containing circle:
area of square, 4
area of circle, π

Square contained in circle:
area of square, 4
area of circle, 2π

Square and circle with equal perimeters:
area of square, π
area of circle, 4
side of square, π
diameter of circle, 4

Square and circle with equal areas:
side of square, $\sqrt{\pi}$
diameter of circle, $\sqrt{4}$

Figure 20. The four relationships between Square and Circle.

diagram shows a square and a circle of equal perimeters. This is the pattern which lies at the foundation of every traditional scheme of sacred geometry. It is known as the circle squared.

Another expression of the circle squared, shown in the fourth diagram, is the square and circle of equal areas. In this, as in all four types of relationship between the square and the circle, the ratio between their dimensions is a variation on $\pi/4$, or about 14/11 or 1.27$\overline{27}$, one of the key ratios in sacred geometry. Its approximation to another common ratio, $\sqrt{\phi}$, is commented on below.

There is no simple way of constructing a square and circle of equal areas, but the squared circle made up of square and circle with equal perimeters can be constructed accurately, if $\pi = 22/7$, by the method shown in figure 21. First, the classic Pythagorean right-angle triangle is drawn with sides in the proportion 3, 4, 5. A square is formed on the shortest side, and a reciprocal 3, 4, 5 triangle is drawn on the opposite side of that square. This creates a base line measuring 11 units, and that line is taken as the side of a square measuring 11×4 or 44 units round its perimeter. A circle is then drawn from the centre of that square so that its circumference passes through the centre of the smaller square which lies between the two 3, 4, 5 triangles. Its diameter measures $11 + 3 = 14$ units, making its circumference equal to $14 \times 22/7 = 44$ units, the same as the perimeter of the square.

This figure is the basic New Jerusalem diagram, and it acquires the

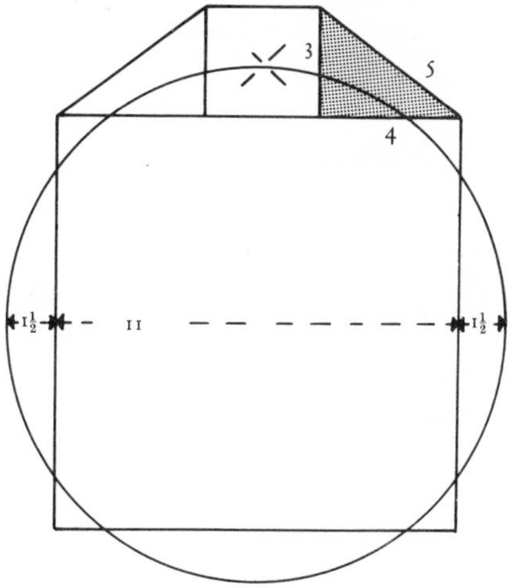

Figure 21. The construction of the squared circle, consisting of a square and a circle with equal perimeters (if $\pi = 22/7$), begins with a Pythagorean right-angle triangle (shaded) with sides in the proportions 3, 4, 5. Multiplied by 720, these base numbers are raised to the dimensions of the New Jerusalem diagram.

appropriate dimensions when the units in which it is measured are multiplied by 720, making the perimeter lengths of both square and circle 720×44 or 31 680 units.

As the square is the geometric symbol of the number 4, so is the triangle the symbol of 3, the pentagon of 5, the hexagon of 6, the heptagon of 7, the octagon of 8 and so on. In the world-image of sacred geometry these and all other numbers find their due place.

The number 3 engenders the first and simplest polygon, the equilateral triangle. It is a progression from the number 2, in that two equal circles, drawn so that each has its centre on the circumference of the other, overlap to create the figure which is the source and womb of the equilateral triangle and of all the other figures of three-fold geometry. The geometer's name for this fish-like shape, which is the area contained by both circles in common, is the Vesica Piscis or Piscine Vessel (figure 23). Within it is the rhombus, consisting of two equilateral triangles. In the symbolism of sacred geometry the two triangles represent the world above and the world below, and other parts of the figure have similar associations, the circle on the right being seen as the realm of spirit penetrating the material world of the left-hand circle, while the longer and shorter axes of the Vesica are respectively positive and negative. The longer axis is equal to the shorter multiplied by the square root of 3, so the ratio between them is $1:\sqrt{3}$ or 1:1.73205 . . . or about 56:97. This means that a square drawn on the longer axis is three times greater in area than the square on the shorter.

A rectangle containing a Vesica is known as a root-three ($\sqrt{3}$) rectangle because of the proportion of its sides. Like all such figures it has an 'eye', situated at the point where a diagonal between two corners is cut by a perpendicular from one of the other corners. This is the eye of the fish in the Vesica. Uniquely, in the case of the root-three rectangle the construction of its eye divides the rectangle into three equal parts (figure 24).

The Vesica Piscis with its three-fold proportions has a special appeal to Christian mystics and is reflected in many symbols, from Jesus the Fish, dominant of the Piscean age, to the shape of a bishop's mitre. Mayananda in *The Wonder Beyond*, a treatise on the Vesica and the esoteric doctrines illustrated by it, says that it was called the Holy of Holies, ὁ ἅγιος τῶν ἁγιων. The value of that phrase is 2368, the number of Jesus Christ.

In the womb of the Vesica the squared circle is conceived. Figure 25 is adapted from a drawing in T.C. Stewart's *The City as an Image of Man*, showing the traditional plan of the Indian cosmic temple. Here the circle and square of approximately equal perimeters are derived in a geometrically pleasing manner from the union between circles.

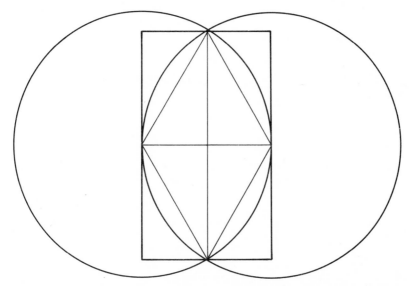

Figure 23. The Vesica Piscis or Vessel of the Fish is the orifice formed by the interpenetration of two equal circles. It gives birth to many of the symbols of sacred geometry. Within it is the rhombus formed by two equilateral triangles, and it is contained in a rectangle with sides in the proportion $1:\sqrt{3}$.

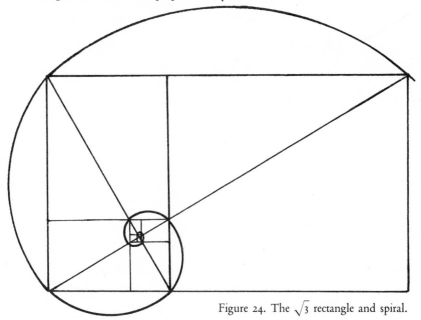

Figure 24. The $\sqrt{3}$ rectangle and spiral.

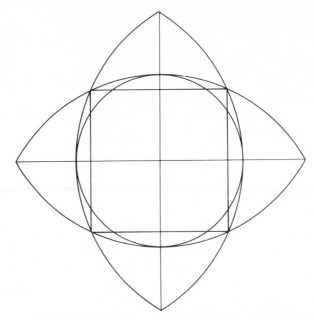

Figure 25. The generative power of the Vesica Piscis extends to the production of the squared circle, which depicts the reconciliation of opposites and is thus the first symbol of the New Jerusalem or divine harmony on earth. In this traditional diagram of temple foundation in India, the double Vesica generates a square and a circle of approximately equal perimeters.

Figure 26. Equal circles pack together as a hexagon.

From the Vesica also comes the hexagon, and the six-fold order of geometry which has several curious and unique characteristics. No one knows just why it should be that the radius of a circle divides its circumference into six equal parts, nor why six circles fit exactly round the circumference of an equal seventh, but so it is; and hexagons are the only regular polygons, apart from squares and equilateral triangles, which pack together leaving no gap between their sides – as illustrated by the cells of a honeycomb and other natural structures. Even more gratifying as a token of geometric order in the universe is the fact that twelve equal spheres can be placed round a thirteenth so that each touches the nucleus and four of its neighbours, producing the geometer's image of twelve disciples grouped round the master. Christ, Osiris and Mohammed are among those who are represented as a central sphere with twelve retainers.

Figure 28. Hexagonal types. *Left*, the two triangles of a hexagram bent into the form of a true lovers' knot, a common symbol in Islamic geometry. *Right*, pattern constructed on a diamond lattice of equilateral triangles.

Pentagonal geometry is entirely bound up with that intriguing ratio known as the golden section, or in mathematicians' shorthand ϕ ($= 1.618\,033\,989...$), which has attracted so much attention from geometers ancient and modern. An example of its wonderful properties is that a rectangle with sides in the proportion $1{:}\phi$ can be divided into two parts, one of them a square and the other a rectangle with the same ϕ proportions (figure 30). The spiral to which it gives rise occurs in many natural forms and patterns of plant growth, giving support to the tenet of traditional philosophy that number preceded creation and determined its development.

The simultaneous construction of a golden section rectangle and a pentagon is shown in figure 31. On a base line AB a square is drawn, and

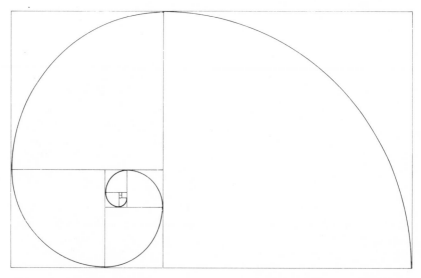

Figure 30. The golden section rectangle with sides in the proportion $1 : \phi$ or $1 : 1.61803 . . .$ increases or diminishes in size by means of the square. When the area of the square on its shorter side is deducted from this figure, the remainder is found to have the same proportions as the original rectangle. The resulting ϕ spiral is one of the common growth patterns in nature.

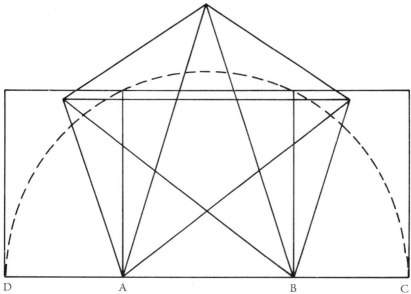

D   A   B   C

Figure 31. A composite of five-fold geometry constructed from the square on base AB. With the pentagon and pentagram is the ϕ or golden section rectangle, longer side AC or DB.

the semi-diagonal of this square is taken as the radius of a semicircle, struck from the centre of the base line and cutting its extensions at C and D. The ratio between the base line, AB, and AC or DB is that of the golden section, $1 : \phi$. The triangle ABE is then formed with its two longer sides, AE and BE, equal to AC and BD, and from that triangle develops the pentagram or five-pointed star, with the pentagon containing it. The golden-section rectangle in this diagram has for its longer side either AC or DB, and its shorter side is one of the sides of the square on AB. The complete rectangle on DC is a root-five rectangle with its sides in the proportion of $1 : \sqrt{5}$ or $1 : 2.2360679 \ldots$ The two smaller rectangles on each side of the square are golden section rectangles. Thus the basic figures of geometry expressing the number five are all displayed in one scheme.

The five-pointed star has been described as a geometer's hymn to the golden section, which frames all its dimensions. It has a particular association with humanity, partly because the proportions of the human body are said to be those of the golden section and partly because of the similarity between a pentagram and a figure with outstretched limbs. It was worn by the Pythagoreans as a talisman of health, and it was also the symbol of their humanistic science. The associated god was Hermes, the medium of revelation and keeper of the Mysteries. Christian mystics made the pentagram an emblem of Jesus, who divided five loaves to feed five thousand people and who represents the archetypal man (with five senses and five fingers to each hand). Its esoteric connection with the crucified man is given further point by the fact that the square on the height of a five pointed star, contained in a pentagon with side measuring one unit, is equal to 2.368 square units, 2368 being the number of Jesus Christ.

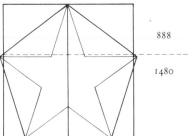

888

1480

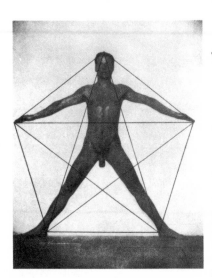

Figure 32. Pentagon formed by the extremities of the human body.

Figure 33. Taking the side of the penta‚ gon to measure 1 unit makes the area of the square drawn on its height equal to 2.368 square units, 2368 being the number by gematria of Jesus Christ. The upper limbs of the inscribed pentagram, symbol of archetypal humanity, nearly coincide with the dotted line which divides the square into two parts, of areas 1.480 (1480 = Christos) and 0.888 (888 = Jesus).

In geometric symbolism the figures deriving from the number five are in direct contrast to those of the number six. Psychic and lunar are the qualities of the pentagon, whereas the hexagon is rational and solar. The dimensions of the solar system are displayed in units of six, and the hexagon is behind inanimate forms such as the snowflake and the crystal. The three angles of an equilateral triangle, six of which make up a hexagon, are each of 60 degrees, reflecting the solar number 666, while between two sides of a pentagon the angle is 108 degrees, 1080 being the characteristic lunar number. An important exercise in sacred geometry is therefore to combine the hexagon and pentagon in one synthetic figure. How this is achieved with tolerable accuracy is illustrated in figure 34. The geometer's myth is that when God created the world he was equipped with nothing but a straight rule and a pair of fixed compasses. With them alone the union of hexagon and pentagon is brought about, the medium of their reconciliation being once again the Vesica Piscis. This beautiful figure was published by Dürer in his *Course in the Art of Measurement with Compasses and Ruler* and is reproduced in C. Bouleau's *The Painter's Secret Geometry*. In the following section (p. 82) is demonstrated how Dürer applied the combined hexagon and pentagon to the composition of his famous *Melencolia*.

There is a strange relationship between ϕ and π which has confused analysts of ancient sacred geometry, particularly in the study of the Great Pyramid. One or other of these ratios dominates all its proportions. To calculate the height of the Pyramid, half the length of its base (378 ft.) is to

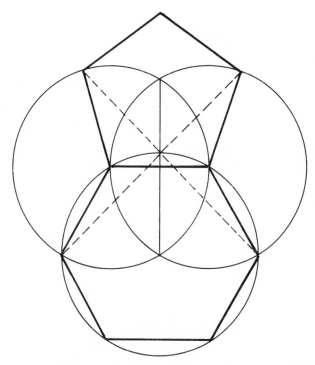

Figure 34. Dürer's classic figure showing the construction of a pentagon from the Vesica Piscis and its union with the hexagon. It is accomplished in the same way as the Creator traditionally designed the universe, 'with the opening of the compasses unchanged'.

be multiplied either by the square root of ϕ or by $4/\pi$. The difference between the two results is so slight that the matter can not be determined by measurement. If π is taken as $22/7$, then $4/\pi = 14/11$, which is virtually the same as $\sqrt{\phi}$. The Pyramid of course was no mere mathematical model, but an example of the cosmic temple whose traditional function was to procure fusion between upper and lower elements. Thus, in imitation of the universe, its designers would have meant to include ϕ, π, and the other key proportions of nature together in one geometric scheme.

Of the other orders of geometry not yet mentioned the most important is the heptagonal. True to character, the number seven, the ungenerated Virgin number, has the most mysterious geometry of all. No one has yet discovered how to draw a mathematically perfect heptagon. It is a secret that, as H.P. Blavatsky put it, 'has not been revealed'. Two approximate methods, both involving the Vesica, are adapted (figures 36, 37) from John James's *Ratio Hunter*.

According to Philo of Alexandria, who was a contemporary of Christ and devoted his life to reconciling the numerical philosophy of his own Jewish people with that of the Greeks and Egyptians, 'Nature delights in the number seven'. He pointed out that there are 7 notes in music, 7 stars in the Great Bear, the astrologers know 7 planets, God rested on the 7th day, a man's head has 7 orifices, his life proceeds in 7-year periods and the cycles of women are synchronized with the 7×4 days of the lunar month. 'By reason of this the Pythagoreans, indulging in myth, liken 7 to the motherless and ever virgin Maiden, because neither was she born of the womb, nor shall she ever bear.'

The mathematical reason for calling seven the Virgin number has been given on page 53, and in geometry the heptagon is also a virgin because it refuses to couple with or generate any other geometric type. Being without birth and unreproductive, the number seven represents the principle of eternity. It is therefore widely represented in symbolism by, for example, the 7 candlesticks, the 7 veils of initiation, the 7 pillars of wisdom, the 7 hills of the imperial city, the 7 colours of the rainbow and the 7 petals of the pomegranate and of the Chinese lotus. The Sumerian temple of the goddess

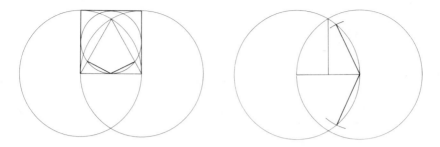

Figures 36, 37. No one has yet discovered how to construct a perfect heptagon; it is one of the mysteries associated with the number seven. Two close approximations, given by John James, have their source in the Vesica Piscis, the matrix figure of sacred geometry.

Left, the square based on the width of the Vesica contains a circle and an equilateral triangle. Lines from the base centre of the triangle form two sides of a heptagon.

Right, half the longer axis of the Vesica is taken as the length of two arcs, struck from the centre of one of the circles forming the Vesica through the circumference of the other circle. Lines from the circle's centre to the points of intersection are two sides of a heptagon.

Nintu at Adab, built some 2500 years before Christ, had 7 gates and 7 doors, and 7×7 animals were sacrificed at its dedication.

St Augustine wrote that '3 is the first number wholly odd and 4 is the first wholly even and these two make 7 . . . . Therefore is the Holy Spirit often called by this number'. He thus admits that to the early Christians the Holy Spirit was a female principle, linked by the number 7 to Pallas Athena. Her name, ἡ ' Ἀθηνα, has the value 77, and Pallas, Παλλας, is one less than 343 or $7 \times 7 \times 7$. Like the spirit of the earth she was influenced by the moon, symbolized by her attendant owl. She was also the Virgin, Parthenos, παρθενος, 515, a number which refers to the heptagon in that the seven-part division of a circle creates an angle of between 51.4 and 51.5 degrees. All this shows how essential it is to include the seven-fold order of geometry in the plan of the cosmological city.

The characteristic New Jerusalem number is outwardly twelve, but esoterically it is seven. Many of the images throughout Revelation are sevenfold. There is mention of 7 churches, 7 kings, 7 angels, 7 mountains, 7 crowns, 7 vials, 7 plagues, 7 thunders, 7 golden candlesticks, 7 stars, 7 lamps, 7 heads and 7 seals, and the four and twenty elders surround a throne on which stands 'a Lamb as it had been slain, having 7 horns and 7 eyes,

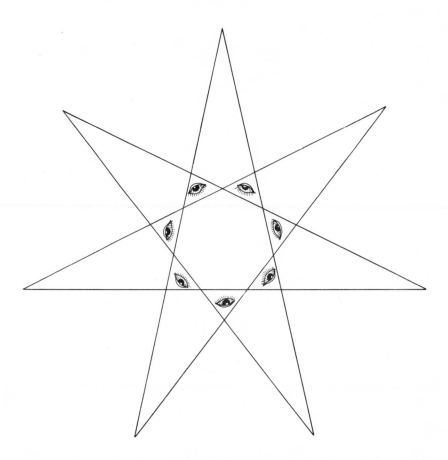

Figure 38. The 'Lamb in the midst of the throne', in Revelation 5,6 has 'seven horns and seven eyes, which are the seven Spirits of God'. The geometer's image of this creature is the heptagram with seven horns and seven triangles containing eyes. The eyed triangle of the heptagon, which has a base angle nearly the same as that of the Great Pyramid in Egypt, may be the origin of the mystic symbol of the eye in the pyramid.

which are the 7 spirits of God'. The seven horns are geometrically equivalent to the seven arms of a heptagram, and in figure 38 the eyes are placed within the seven small triangles within the figure. Herein may be seen the origin of that most famous esoteric symbol, the eye within the pyramid. It is commonly supposed to represent the Great Pyramid at Giza with its base angle of some 51° 51', but the angle here, though almost the same, is that of a heptagon, some 51° 26'.

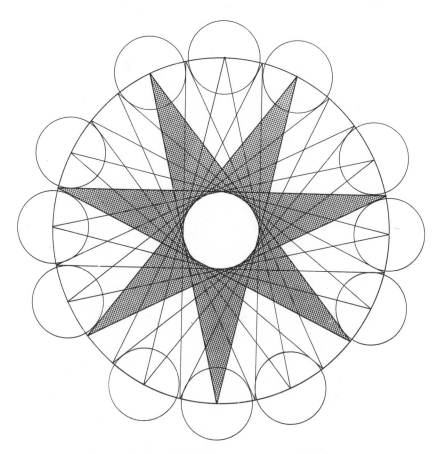

Figure 39. References throughout the chapters of Revelation to the geometry of the New Jerusalem repeatedly demand that the number 12 be combined with the number 7 to symbolize the union of body and spirit. This union is achieved through the New Jerusalem ring of twelve lunar circles, arranged as in Figure 9. Accommodated by this ring is a figure made up of four heptagrams, having 28 horns, the number of phases in a lunar cycle. The regularly spaced horns fit neatly between the circles, touching their sides, or terminate at their centres.

Four heptagrams with a total of 28 'horns' are accommodated in the New Jerusalem plan as shown in figure 39. They are contained by the circle of circumference 31 680 which is seen to divide naturally into 28 equal parts. The number 28 gives the days in the lunar month and the female cycle, while the heptagrams, figures of the Virgin number seven, represent the Bride in the symbolism of New Jerusalem geometry.

Sacred geometry in Dürer's 'Melencolia'

In 1979 the artist Franz Deckwitz of Amsterdam presented the author with a copy of his magnificent folio work, *Dürer's Melencolia with Compass and Ruler*, which he published in an edition of 150 copies. It contains his interpretation of the hidden geometry in the design of Dürer's famous engraving of 1514, depicting a winged, dark-visaged, brooding figure, seated amid apparatus of geometry and masonic crafts. The picture is crammed with symbolism, and the question of what Dürer meant by it has given rise to a vast literature. Deckwitz's approach, however, has rarely been tried. His pioneering efforts open the way to a fuller understanding of *Melencolia* through the symbolism of the geometric types underlying it. Building on his work, it becomes possible to establish with some confidence the nature of Dürer's geometric and numerical statement.

The first part of the enquiry concerns the unit of measure by which Dürer worked. This proves to be a fraction of the shorter Roman foot as defined in the next chapter (page 96). Its length is 0.967 68 ft. For the purposes of printers and designers the foot was divided into 12 inches and the inch into 12 lines or 72 points. The foot therefore consisted of 864 points. The particular unit employed by Dürer in *Melencolia* appears to have been one of 100 Roman points or 1.344 English inches (3.413 767 centimetres).

The primary evidence for this unit is shown in Figure 41. Most of the shapes and instruments in the design are displayed at an angle, foreshortened, and can not therefore yield a true measure. Exceptions are the sphere, bottom left, and the measuring stick, bottom right, which lies parallel to the margin. The sphere is the most clearly defined shape, and its diameter measures 1.344 inches. This is the unit here referred to as the Dürer inch (D''). The measuring stick is broken and jagged at each end, so its length is indeterminable; but its important features are its two 'eyes' which were made to receive pegs defining a precise length. Between the two eyes (one of which contains a white dot, the exact point of measure) the distance is 3.36 inches or $2\frac{1}{2}$ D''. In classical metrology a unit of $2\frac{1}{2}$ feet was called a pace, and it is reasonable to suppose that the measuring stick depicted in *Melencolia* was a Roman pace measure and that its actual length between the eye holes was $2\frac{1}{2}$ shorter Roman feet or 2.4192 ft. In that case the D'' unit represents one of these feet and the scale of the design is established as 100:864.

Also in conformity with the D'' unit is the height of the print, the published length of which is 23.9 centimetres = 9.408 inches = 7 D''. The

Figure 40. Dürer's *Melencolia* of 1514. The date appears in the two central panels at the bottom of the magic square above the seated figure. One inch on the reproduction corresponds to one 'Dürer inch' (D") on the original.

width of the print is just under $5\frac{1}{2}$ D", but Deckwitz has pointed out that the thin margin between the outline of the print and the copper plate on which it was engraved plays a significant part in the underlying geometry. Measured between the centres of the print margin, the width is $5\frac{1}{2}$ D" exactly. This allows the drawing of a rectangle 7 D" high and 5.5 D" wide (figure 42), and the importance of the ratio between the two sides (11:14 or 1:1.27$\overline{27}$) is that it produces the 'squared circle'. A circle with diameter equal to the height of the print, 7 D", has a circumference of 22 D", and that is also the length of the perimeter of the square with side measuring 5.5 D".

Figure 42 shows the square and circle of equal perimeters which derive from the proportions of *Melencolia*, together with a pentagon which features in the hidden geometry of the design. The significance of this pentagon in relation to the general scheme of geometry behind *Melencolia* is seen in figure 43. Each of its five sides gives the appropriate pentagonal measure, 5 D″, so the length of its perimeter is 25 D″, the same length as the perimeter of the 7 × 5.5 D″ rectangle which frames the print. The four figures, square, circle, pentagon and rectangle, outlined in Figure 42, are interrelated proportionally in a number of ways, as detailed beneath the figure.

By the most obvious of clues Dürer makes it plain that the principal geometric theme in *Melencolia* is pentagonal. The legs of the ladder, one of which meets the wall of the house at the exact top centre point of the composition, slope at an angle of 18°, thus indicating a pentagram or five-pointed star. This is in accordance with Dürer's habit, apparent in many of his designs, of emphasizing a top-centrally placed angle to typify the underlying geometry. In figure 43 a pentagon is drawn within the top margin of the print, its size is determined by the line of the horizon, top left, and it is found to stand on a square based on the lower margin. The side of that square is equal to half the width of the print. A circle is drawn to enclose it, and in that circle a hexagon is drawn. The hexagon's upper side

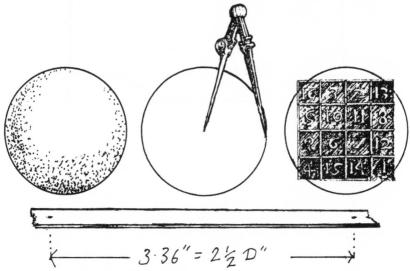

$$3 \cdot 36'' = 2\tfrac{1}{2}\,D''$$

Figure 41. Items in Dürer's *Melencolia* displaying his unit of measure, 1.344″ (1D″).

diameter of sphere $= 1\mathrm{D}''$
opening of compass $= \tfrac{1}{2}\mathrm{D}''$
distance between eyes of ruler $= 2\tfrac{1}{2}\mathrm{D}''$

One of the squared circle relationships in *Melencolia*, noticed by Deckwitz, is formed by the magic square and the sphere. Placed together, above, they are found to have approximately equal perimeters.

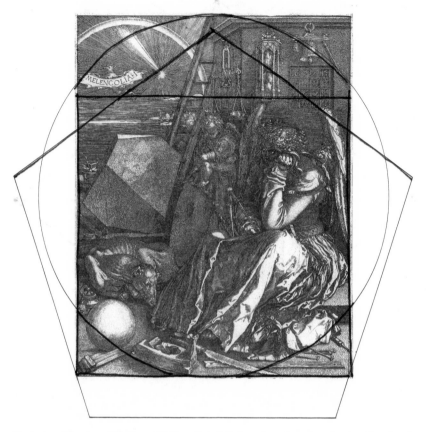

Figure 42. The overall design of *Melencolia* is based on four shapes proportionally related: square, circle, pentagon and 'squared circle' rectangle. The longer side of the rectangle, equal to the height of the print, is 7D″, and its width, measured from the points between the margin of the print and the edge of the plate, is 5½D″.

The area of this rectangle is equal to the area of the circle with perimeter of 7D″, the height of the print. The circumference of that circle, 22D″, is equal to the perimeter of the square on the shorter side of the rectangle.

Two sides of the pentagon cut the corners of the rectangle, and another is tangent to the sphere shown in the picture. The measured length of its side is 5D″, so its perimeter is 25D″, equal to the perimeter of the rectangle. Thus:

$$
\begin{aligned}
\text{area of rectangle} \quad &= \text{area of circle} \quad &&= 38.5 \text{ sq. } D'' \\
\text{perimeter of rectangle} &= \text{perimeter of pentagon} &&= 25 D'' \\
\text{perimeter of square} \quad &= \text{circumference of circle} &&= 22 D''
\end{aligned}
$$

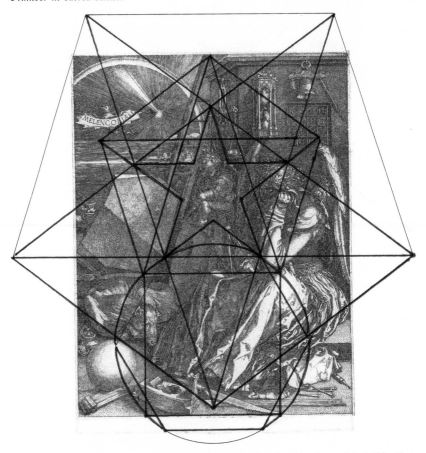

Figure 43. The hidden scheme of geometry behind *Melencolia*. The slope of the ladder from the top centre of the print gives the pentagram angle, and so does the line from the top left which passes through the point on the sea horizon where Dürer rested the point of his compass to draw the arc of the rainbow, then down the axis of the geometric solid and through the eye in the ruler. The completed figure contains Dürer's construction (Figure 34) of the union of hexagon and pentagon. Contained within the central pentagram is the winged child of inspiration. By this synthesis of geometric symbols is expressed the inner meaning of *Melencolia*.

coincides with the base of the small pentagon which stands at the centre of the greater five-sided figure. Beneath the surface of *Melencolia* are thus discovered two of the classic constructions of sacred geometry, the squaring of the circle and the combination of hexagon and pentagon.

The hidden scheme of geometry which emerges from this analysis is firmly based, as far as it goes, on the clues which Dürer concealed in the details of his composition. It is by no means offered as a complete interpretation of the pattern below the surface of *Melencolia*, but it sufficiently demonstrates the main features in the construction. The artist has combined together various symbolic figures within a scheme governed by the pentagon and the number 5. That number is evidently the key to the esoteric meaning of this strange work of art.

The five-part division of the circle by the five extremities of the body (figure 32) illustrates the traditional meaning of the pentagram and other figures of five-fold geometry as emblems of humanity and the microcosmos. They pertain also to Hermes, who represents the principle of inspiration in art and science and is therefore called the messenger of the gods. The pentagram has been used to symbolize initiation into the mysteries, the Pythagorean secret science and the esoteric teachings of Jesus. It is certainly appropriate to the subject of *Melencolia*. Scholars have proved that Dürer's reference was to a passage in the *Occult Philosophy* by his contemporary, H. Cornelius Agrippa, describing the 'melancholy humour' which visits artists, poets and all those who engage throughout the hours of darkness in striving to unveil the secrets of nature. It is the melancholy of genius, a state in which the solitary, devoted student is possessed by the daemons of inspiration. In that state, says Agrippa, knowledge and wisdom are acquired. Philosophers at the Renaissance were deeply interested in the melancholy humour, the advantages to be gained by invoking it and the price to be paid in terms of health and spirit. Ficino (quoted in Frances Yates's book *The Occult Philosophy*) advised that the dark Saturnine humour which comes with obsessive study should be mitigated by Jovial or Venereal influences. Attracting the powers of Jove is the function of the magic square of Jupiter, which in *Melencolia* is placed-above the brooding figure. Yet the figure is by no means jovial. Wings folded, listlessly twirling a compass, she appears sunk in despair. Around her are strewn the instruments which her studies of the universe have caused her to acquire. Evidently these have served their purpose, and the figure seeks a higher stage of initiation. Thus she directs her eyes upwards, towards the cherub who sits at the centre of the hidden scheme of geometry, and to the comet and rainbow behind him. The figure's melancholy trance is induced and deliberate, for the acquisition of learning. She has passed through the first stage of initiation, having mastered the crafts and practical sciences. Now she is ready for higher forms of study.

Inspiration, says Agrippa, 'occurs in three different forms corresponding to the threefold capacity of our soul, namely the imaginative, the rational and the mental'. In its imaginative function the soul 'becomes a habitation for the lower daemons, from whom it often receives wonderful instruction in the manual arts'. This type of possession is evidently depicted in Dürer's print. Its full title, *Melencolia I*, suggests that Dürer planned later to illustrate Agrippa's other two categories. The next of these relates to the rational part of the soul, exposing it to the middle sort of daemons who inspire knowledge of nature and humanity and instruct in philosophy. The third form of possession, the mental or intellectual, is by the higher spirits, inducing direct knowledge of divinity, the eternal laws of the universe, the nature of the soul and the destiny of all things.

Dürer was known in his time as a profound scholar and the most brilliant exponent through his art of the ancient philosophy which inspired the Renaissance. In the spirit of humanism he investigated the proportions of the human body as reflections of the mathematically ordered universe. His particular medium for depicting, and thus magically procuring, the union between the human microcosm and the divine macrocosm was sacred geometry, of which he became the greatest master. Beneath the figures, shapes and objects on the surface of *Melencolia* there lies Dürer's most valuable gift to the aspiring initiate, a symbolic design of pure geometry, an ideal framework to sustain the beleaguered mind. Dominating it is the five-pointed star to invoke the mood for scholarship and revealed knowledge. Wedded to the central pentagram in figure 43 is the hexagon, their two figures combining to symbolize the purpose of a true scholar, to relate his own interests to those of the world at large. Another pentagram in the design rests firmly on a square, signifying perhaps the imaginative part of the mind placed on solid, four-square foundations. Also displayed in *Melencolia*, both superficially and in the hidden geometry, is the figure of the squared circle. As a symbol of reconciliation between different orders in nature, its presence here prepares the student of life's mysteries for the contradictions and paradoxes he will encounter and provides the geometric model to contain them.

The theme of *Melencolia*, as here interpreted, is the mind of the mystic and its ascent through studies of the worldly sciences towards higher forms of contemplation and wisdom. Dürer's work is both an illustration and a guide to that process. The lonely scholar attracts the melancholy humour and the spirit of revelation together, and at such times he is vulnerable to obsessions, delusions, and psychic attacks, possibly with long-lasting effects. *Melencolia* is designed as his talisman, for the pattern on which it is based represents the principle of inspiration combined with order and sanity. Thus provided is a mental image of the true philosopher, who passes unscathed through periods of creative frenzy and mania, preserved from ill effects by an educated and ever-vigilant sense of proportion.

Music and proportion

Plato's theory of education was that children must be exposed from their earliest days to the influence of harmonious proportions in everything around them, so that they will grow up with a sense of proportion and the ability to distinguish between the good and the meretricious. The strongest of all influences is music. That is why in ancient societies, and in Plato's prescription, its forms were strictly controlled. The soul of every child is naturally attuned to the harmonies of the universe which, said the Pythagoreans, arise from the notes emitted by the planets in the course of their perfectly ordered circuits, inaudible to us because our ears are accustomed to them. As the planets revolve their song changes, allowing different tones to prevail, which explains why at different times there are different fashions in music, but its underlying harmony remains the same. The intervals of music can be expressed as ratios between numbers, so the composition of the world-soul was resolved into a numerical canon, from which certain musical scales were derived, the only ones considered lawful. By forbidding 'noise', as non-canonical music was called, Plato and his ancient forebears intended to preserve the soul from disruptive influences, nurture it on those sounds which are conducive to its natural development and thus produce the type of citizens who will appreciate and maintain a society constituted in imitation of the cosmos.

In the *Republic* Plato wrote derisively about students of music who 'torment catgut and try to wring the truth out of it by twisting it on pegs' but 'never get as far as formulating problems and asking which numerical relations are concordant and why'; and he was equally scornful of astronomers who looked upwards at the sky, 'in the hope of learning the truth about proportions from the actual movements of the heavenly bodies'. The planetary system, he allowed, was very beautiful because it was the work of the greatest of all artists, but its material aspect was only a reflection of its true nature. The philosophic astronomer should therefore 'look down, not up', for the secrets of the universe are not to be found in phenomena but in the precise proportions of number. Similarly in music, the ear can detect harmonies only approximately, so it is an inadequate instrument for studying their actual structure. The only way they can properly be understood is through numerical analysis of their proportions. That study, according to Plato, is of greater benefit than the actual practice of music.

The science of relating music to number is far older than history. Pythagoras, who revived it and thereby instigated a renaissance in Greek and western culture, discovered it from his studies of traditional Babylonian science and then proceeded to test its basic principles. He or his followers experimented with stretched strings, water jugs, flutes of different lengths and in many other ways in order to show that the fixed tones in music corresponded to simple number proportions. Two strings, one twice the length of the other, were found when plucked simultaneously to sound the octave (1:2), and the same result was obtained from two vessels, one empty and the other half full of liquid. The proportions 4:3 and 3:2 produced the two parts of the octave, the fourth and the fifth, and 9:8 gave the interval between two notes in the scale, the tone. The numbers 1, 2, 3, 4, which generate by addition the Tetractys (see p. 51), could be combined in different ways to make the octave (1:2 or 2:4), the double octave (1:4) and the fifth (2:3), and a further series, 6, 8, 9, 12, yielded all the fixed tones of the musical scale: the octave (6:12), the fourth (6:8), and fifth (6:9) and the tone (8:9).

In the numbers 6, 8, 9, 12 were also found two of the three types of proportions which the Pythagoreans recognized between any two numbers, those three being the arithmetic, the geometric and the harmonic. In arithmetic proportion the mean term is equal to half the sum of the extremes and is thus placed half way between them, as in the series 6, 9, 12. In geometric proportion the mean is to the first term as the third term is to the mean, as in the series 4, 6, 9, where $4 \times 1\frac{1}{2} = 6$ and $6 \times 1\frac{1}{2} = 9$. In the case of harmonic proportion the mean exceeds the smallest term and is exceeded by the largest by the same fraction of the extremes. An example is 6, 8, 12, where 8 is equal to 6 plus its third part and 12 minus its third part. The formula for finding the arithmetic mean between two numbers, a and b, is $(a + b)/2$, that for the geometric mean is $\sqrt{ab}$, and for the harmonic mean $2ab/(a + b)$.

The earliest known account of these three types of proportion and the part they play in Pythagorean music and number theory is in the *Timaeus*, where Plato describes the numerical creation of the physical universe and the soul which binds and animates it. The conventions of symbolic arithmetic forbid the use of fractions because the character of the monad is to be indivisible, so if fractions occur in a musical scale or progression of numbers all the terms in it must be multiplied by their lowest common denominator to make them all integers. The more extensive a scale, the higher become the numbers needed to express it. Thus Plato attributed to the world-soul a range of over four octaves in order to bring out such numbers as those listed on page 54, constituting the number code which was at the root of his philosophy. Over many centuries scarcely any two scholars have agreed on the exact series of numbers which made up the world-soul, but it has been commonly accepted that it included the majority of those given on that page, where the matter is more fully explored.

The numbers of the world-soul which occur naturally as intervals in music were adapted to units of measurement, so that the sounds produced by wind or stringed instruments corresponded numerically to the length of a pipe or a string. This science indeed went far beyond music, for the same numbers and units of measure were found also to apply to astronomy and the measurement of time. In this study it is impossible to locate origins; the code of number behind ancient science appears seamless and, like the circumference of a circle, with no beginning. So many different and seemingly unrelated classes of phenomena are subservient to it that it would be invidious to identify any one as the source and primal reference of the symbolic numbers here examined. One is drawn therefore into sympathy with the mystical view constantly urged by Plato (who was impelled to mysticism by science rather than predisposition), that number is behind all things and is the bond which holds together every diverse form of nature.

A circle of perpetual choirs

One of the instructions Plato lays upon his Magnesian colonists (see page 167) is that they should seek out the shrines of the native gods and spirits, perpetuate their cults and hold festivals at their sites. There should, he says, be no less than 365 festivals in the country during the course of the year. This idea of an unbroken festive cycle is matched by his belief in choral singing as a beneficial exercise for all citizens. The modes of music he prescribed for them were canonical, echoing the harmonies of the Heavenly Choir as described at the end of the *Republic*.

Plato was a respecter of tradition, a revivalist rather than an innovator, and the important themes in his writing were drawn from his knowledge of ancient science and sacred institutions of the past. His emphasis on choirs harks back to the days of priestly rule, when the calendar was regulated by the temple authorities and the procession of seasons and cycles was marked by an endless round of chanting and ceremonies. As the times changed so did the music of the temple, reflecting the movements of the planets. Their relative positions at any time determined the prevailing musical mode. At certain intervals, judged significant by the astrologers, when the planets repeated a particular formation, one cycle of temple music would come to an end and another begin. In response to this, subtle changes would become evident in religious symbolism and the forms of society. Thus the fluctuating moods of human nature were each allowed regular expression, in the course of a year and over greater periods of time. The institution of the temple was an attempt to perpetuate those legendary days when government was conducted through the influence of music rather than by means of a rigid code of law.

A relic of those days may be seen in the tradition of the Perpetual Choirs of Britain. In the Welsh Triads, verses which are thought to incorporate elements from ancient bardic lore, the sites of three of the Perpetual Choirs are named as Glastonbury, the Choir of Ambrosius or Stonehenge, and Llan Illtud Fawr which is the old Celtic sanctuary at Llantwit Major in Glamorgan. At each of the Choirs 2400 saints maintained a ceaseless chant, 100 for every hour of the day and night. Similarly, in Revelation, 24 elders stand before the throne of the Lamb, 'every one of them having harps, and golden vials full of odours, which are the prayers of saints. And they sung a new song . . .'. The elders' new song was their accompaniment to the appearance of the New Jerusalem and a new dispensation on earth.

There is a curious symmetry about the positioning of the three Perpetual Choirs in Britain. Stonehenge and Llantwit Major are equidistant from Glastonbury, some 38.9 miles away, and two straight lines drawn on the map from Glastonbury to the other two Choirs form an angle of 144°. This same angle also occurs at Stonehenge. The axis of Glastonbury Abbey points towards Stonehenge, and there is some evidence that it was built on a stretch of ancient trackway which once ran between the two Choirs. Its line from Glastonbury is orientated about 4° north of due east, and the avenue at Stonehenge extends the axis of the monument towards midsummer sunrise in a direction 50° east of north. Thus the angle at Stonehenge between the line from Glastonbury and the sunrise line is close to 144°. If the sunrise line is extended 38.9 miles from Stonehenge it terminates at the site of another

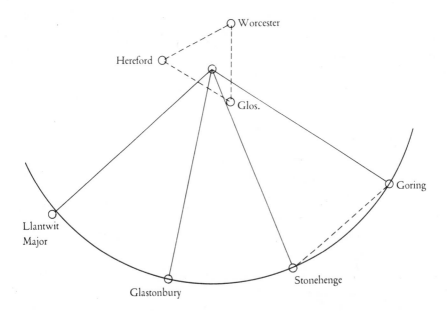

Figure 44. The English cathedrals of Worcester, Gloucester and Hereford, where the modern Three Choirs Festivals are held, are almost equidistant from each other, thus forming an equal-sided triangle near the centre of which is the Whiteleafed Oak where the three counties meet. That spot is the secluded centre of the archaic Circle of Perpetual Choirs by which the Enchantment of Britain was formerly maintained. Three of the choirs were located at Stonehenge, at Glastonbury and near Llantwit Major in Wales. Others appear to have been at Goring-on-Thames and at Croft Hill in Leicestershire, a traditional site of ritual, legal and popular assemblies.

sanctuary, a temple which formerly existed at Goring-on-Thames, where several prehistoric tracks converge at a river crossing.

The number 144 is characteristic of the New Jerusalem scheme, and 144° is the angle between two sides of a decagon. Llantwit, Glastonbury, Stonehenge and Goring therefore occupy four consecutive corners of a regular ten-sided figure, the centrepoint of which proves to be a spot called Whiteleafed Oak which is the meeting-place between three counties, Hereford, Worcester and Gloucester. It was once the site of a sacred grove with Druidic associations, which local people held in awe even in the nineteenth century. It is in a valley of the Malvern Hills, said to be the oldest geological formation in Britain, and just above it is the prehistoric sanctuary on Midsummer Hill. Writing about the Whiteleafed Oak in his book of

1875, *The British Camp on the Herefordshire Beacon*, James McKay recorded that 'within the memory of living men an oak existed . . . round which superstitions still lingered, which were unquestionably of Druidical origin. Its leaves were dotted over with blotches of white . . . and their size, and arrangement, and number were taken as so many omens by the credulous rustic, of the fortunes, good or bad, which the Fates had in store for him'.

Its vague but old-established reputation as a sacred spot and its position on the three-county junction suggest that the Whiteleafed Oak was a significant point in the sacred geography of archaic Britain. And it is delightfully appropriate that it should be found to stand at the hub of the circle of Perpetual Choirs, for at the cathedrals of Hereford, Worcester and Gloucester, the counties which meet together at the Whiteleafed Oak, are held today the famous musical festivals of the Three Choirs.

The circle of Perpetual Choirs, centred on the Whiteleafed Oak, has a radius of 63 miles or 504 furlongs – a measure which relates it to the New Jerusalem circle of radius 5040 ft. and circumference of 31 680 ft. The scale ratio between the two figures is 10 sq. ft. to 1 acre, for the area of the New Jerusalem circle is 79 833 600 or $2 \times 11!$ sq. ft. and the circle of Perpetual Choirs contains 7 983 360 acres.

The circle of Perpetual Choirs must be older than any structure existing today, older even than the present Stonehenge which was built about four thousand years ago on a site of far earlier sanctity. Both the plan of Magnesia and the New Jerusalem foundation-plan at Glastonbury can be seen as restorations, designed to reinvigorate the tradition of sacred science; and it may be that Stonehenge had the same purpose. Here one is limited to speculation, but it is at least conceivable that when its builders erected the Stonehenge sarsen circle with its mean radius of 50.4 ft., they were aware that this dimension reflected the 504-furlong radius of the Perpetual Choir circle on a scale of 1 : 6600; and they may have constructed their temple as part of a revival programme, aimed at restoring the musical enchantment with which the ancient bards held a whole country in harmony with the heavens.

3 Number and measure

TRADITONAL AND ANCIENT units of measure are essentially but numbers. They are all indeed fractions of the earth's dimensions, but that reference is secondary, for both the dimensions of the earth and the values of the ancient units of metrology derive in the first instance from that same code of number as occurs in all categories of natural phenomena. The ancient units are in harmony with the measurements of the earth, not simply because they were so designed, but because both systems are reflections of the one numerical standard.

There are two ways of approaching the problem of ancient metrology, one physical and imprecise, the other rational and exact. The first is through analysis of the most careful measurements taken by archaeologists of structures such as the Parthenon, Stonehenge and the Great Pyramid of Cheops. The second is through number, and the results in both cases are the same.

The measurement of the Great Pyramid has occasioned much acrimonious debate among mystics, British Israelites and Egyptologists, and the amount of nonsense written on all sides has effectively obscured the subject. Yet statements by Greek and Roman authors, including Herodotus, Agatharchides and Pliny, show that the Pyramid's dimensions were considered significant in their time, and the meticulous survey of the Pyramid for the Egyptian Government by J.H. Cole in 1926 has provided the data for recovering the measures used by its builders.

Cole reported that the average length at the base of each of the Pyramid's four sides was 755.785 ft. This is within about $2\frac{1}{2}$ inches of the length evidently intended, 756 ft., which is equal to 500 cubits of 1.512 ft. A cubit of this length has been identified from the measurements of other ancient monuments and is generally accepted by metrologists. It is the unit referred to by Agatharchides who wrote that the side of the Pyramid measured 500 cubits. The corresponding foot (two-thirds of a cubit) is equal to 1.008 ft. These units are called the shorter Greek cubit and the shorter Greek foot.

It is recorded in early literature that 24 Greek feet or cubits were equal to 25

Roman. The shorter Roman foot was therefore of 0.96768 ft., and 3125 of these feet made up the 3024-ft. distance round the four sides of the Pyramid's base. Agatharchides also mentions a stade, of which there were 5 in the perimeter of the Pyramid. This stade was therefore of 600 Greek feet, equal to 625 Roman feet or 604.8 ft.

An important clue to the ratios of ancient metrology was provided by two old Greek writers, Epiphanius and Hesychius, who stated that seven stades composed of Egyptian feet were the same as the Roman mile. The Roman mile was of 5,000 Roman feet, so it was equal to 4838.4 ft. This makes the length of the Egyptian stade 691.2 ft., and the Egyptian foot, taking 600 feet to the stade, was therefore of 1.152 ft. The corresponding cubit is of 1.728 ft., and is known as the Egyptian royal cubit. There are 1750 of these cubits in the Pyramid's base perimeter.

However, from measurement of the Pyramid's internal chambers, archaeologists have concluded that a shorter version of the royal cubit was used and that there were 1760 of those shorter cubits in the Pyramid's perimeter or 440 in each of its sides. The shorter cubit therefore relates to the longer version as 175:176, making it equal to $1.718\overline{18}$ ft. The corresponding foot is of $1.145\overline{45}$ ft. In this value is a reference to the $4/\pi$ or $14/11$ ratio which dominates the proportions of the squared circle (see pp. 68–69), for $14/11 \times 9 = 11.45\overline{45}$.

This ratio of 175:176 evidently obtained within the other units of ancient metrology, for F.C. Penrose, who surveyed the Parthenon towards the end of the nineteenth century, found that its rectangular platform measured 100 by 225 Greek feet of length 1.01379 ft., and if that is minutely adjusted to 1.01376 ft., it relates as 176:175 to the shorter Greek foot of 1.008 ft. Virtually the same 176:175 ratio between the longer and the shorter Roman feet was recorded by the pioneer metrologist, Professor John Greaves, from his investigation of Roman monuments in the seventeenth century. The existence of two versions to each of the old units was accepted by the middle of the nineteenth century when John Taylor, in his work on ancient metrology *The Great Pyramid*, declared that 'there were two Greek feet, which held the same relation to each other that the two Roman feet appear to have done'. He referred to the shorter Greek foot, which Herodotus called the foot of Samos, as the Ptolomaic foot.

The basic reason for two closely related versions of the same units becomes clear on investigation. It has to do with number symbolism. In ancient science, symbolism was always wedded to practicality, and the useful purpose of the longer and shorter versions of each unit was to provide two different lengths for the nautical mile, which is equal to one minute in a degree of latitude along the earth's meridian. The earth's deviation from the form of a true sphere means that the length of each successive degree of latitude increases from the equator towards the poles. At about 10° latitude north or south the length of a degree is 362 880 (= 9!) ft., making the length of a minute of arc 6048 ft. This is equal to 10 stades of 604.8 ft., of which there are 5 in the perimeter of the Pyramid.

At about 50° N or S the length of the nautical mile has increased to 6082.56 ft., approximately the value used by modern navies. It is equal to 6000 longer Greek feet of 1.013 76 ft. or to 6250 longer Roman feet of 0.973 209 6 ft. or to 5280 longer Egyptian feet of 1.152 ft. The shorter nautical mile of 6048 ft. is made up of 6000 shorter Greek or 6250 shorter Roman or 5280 shorter Egyptian feet, and it is also divided by their respective cubits and stades. The difference between the longer and shorter nautical miles is 34.56 ft. or 30 Egyptian feet of 1.152 ft.

Multiplying the length of the average nautical mile, as indicated by ancient metrology (6082.56 ft.) by 21 600, the number of minutes in a circle, produces the traditional value of the length of the meridian, the great circle passing through the poles. It proves to be 131 383 296 ft. or 24 883.2 miles, the mathematical significance of the mile measure being that $248 832 = 12^5$. The units which make up this distance, as listed below, do so by multiples which are those duodecimal numbers common to all ancient sciences.

131 383 296 ft. $= 12^6 \times 44$ ft.

$= 135 000 000$ Roman feet of 0.973 209 6 ft.

$= 90 000 000$ Roman cubits of 1.459 814 4 ft.

$= 216 000$ Roman furlongs of 608.256 ft.

$= 27 000$ Roman miles of 4866.048 ft.

$= 129 600 000$ Greek feet of 1.013 76 ft.

$= 86 400 000$ Greek cubits of 1.520 64 ft.

$= 207 360$ Greek furlongs of 633.6 ft.

$= 25 920$ Greek miles of 5068.8 ft.

$= 114 048 000$ Egyptian feet of 1.152 ft.

$= 76 032 000$ Egyptian cubits of 1.728 ft.

The length of $12^5/10$ miles is decidedly closer to modern estimates from satellite data of the earth's meridian than the estimate made of that distance by the French scientists in the eighteenth century for the purpose of defining the metre. The metrologist A.E. Berriman writes that, had the French accepted Cassini's proposal in 1720 for a scientific foot based on a six-thousandth part of a minute of average latitude, that unit would later have been recognized as identical with the longer Greek foot by which the Parthenon was built.

Lengths of ancient units of measure

The names, Greek, Roman, Egyptian etc. are applied to these units by convention only, for they belong to the same number system and represent fractions of the earth's dimensions. Their lengths are given here in English feet (ft.):

		longer values	shorter values
Roman	foot	0.973 209 6	0.967 68
	cubit	1.459 814 4	1.451 52
Greek	foot	1.013 76	1.008
	cubit	1.520 64	1.512
Egyptian	foot	1.152	1.145 $\overline{45}$
	cubit	1.728	1.718 $\overline{18}$

ratios: Greek : Roman = 25:24
Egyptian : Greek = 25:22
longer : shorter = 176:175

The values of the units listed above are the same as those which measure Stonehenge. The structure and dimensions of that very ancient monument are described more fully on pages 29–34. Its main feature is the sarsen circle, the once unbroken ring of 30 arc-shaped lintel stones elevated on 30 pillars. The published figure for the inner diameter of the sarsen circle is 97.32 ft. This being taken as 97.320 96 ft., it is equal to 100 longer Roman feet or 96 longer Greek feet at the values given above. The mean diameter of the sarsen circle, measured to the centre of the stone lintel ring, is 100.8 ft. or 100 shorter Greek feet, and its circumference, 316.8 ft. (a hundredth part of 6 miles) is equal to 3125 longer Greek feet.

The powers of 12 are the agents by which the terms in the π fraction, 22/7, are raised to the wonderful numbers of the Stonehenge or New Jerusalem circle and then outwards to give the canonical dimensions of the earth.

$22 \times 12^2 = 3168 = $ circumference

$7 \times 12^2 = 1008 = $ diameter

$3168 \times 12^4 \times 2 = 131{,}383{,}296 = $ earth's meridian in feet

$1008 \times 12^4 = 20{,}901{,}888 = $ earth's mean radius in feet.

The most important order of metrology in the dimensions of the Stonehenge circle is that known as Hebrew, which is referred to in Biblical accounts of the Temple at Jerusalem. The tradition was that its units represented fractions of the earth's polar axis, and for that reason it was investigated by Sir Isaac Newton during his inquiry into the dimensions of the earth as reckoned by the ancients. His researches led him to attribute a length approximating to 2.0736 ft. to the unit he called a sacred cubit which is really a double foot. This figure was made exact by John Taylor, who recognized it as equal to six-fifths of the longer Egyptian cubit of 1.728 ft. A unit of 2.0736 ft. is, however, considerably too short to represent a ten-millionth part of the earth's polar radius, so Taylor supposed the existence of a larger version. Multiplying 2.0736 ft. by the 176/175 fraction which expresses the two different versions of the other units produces a unit of 2.085 449 1 ft., and that unit is indeed a ten-millionth of the polar radius. The radius is thus made equal to 3949.7142 miles, which is virtually identical to modern estimates of its length.

The distance of 20 854 491 ft. or $2^4 \times 12^3/7$ miles, being the shortest from the centre of the earth to its surface, relates to the meridian circumference of $12^5/10$ miles, not as $1:2\pi$, but as 10:63.

The polar radius was also divided into 12 and 6 million parts, its twelve-millionth part being the cubit-and-a-handsbreadth of the Old Testament. The handsbreadth was a seventh part of a cubit, and the unit known as a cubit-and-a-handsbreadth was equal to the longer so-called Greek cubit plus its seventh part, or 1.737 874 ft. Also, it is in geometric proportion with the longer and shorter versions of the Egyptian cubit, 1.728 ft. and 1.718 $\overline{18}$ ft., for

1.737 874 $\times$ 175/176 = 1.728, and

1.728 $\quad\times$ 175/176 = 1.718 $\overline{18}$

The double cubit-and-a-handsbreadth or rod of 3.475 748 5 ft. represented a six-millionth part of the polar radius, and it is evident from a passage in Pliny that the radius was also divided into 21 000 000 parts by a 'polar foot'.

Referred to generically as Hebrew or sacred units are those which apply particularly to the earth's polar axis. Those already mentioned are:

Polar radius = 20 854 491 ft.
            = 21 000 000 polar feet of 0.993 071 02 ft.
            = 20 000 000 Hebrew feet of 1.042 724 55 ft.
            = 12 000 000 cubit-and-a-handsbreadths of 1.737874 ft.
            = 10 000 000 Hebrew double feet of 2.085 449 1 ft.
            = 6 000 000 sacred rods of 3.475 748 5 ft.

The so-called Hebrew stade of 500 Hebrew feet was the unit by which Eratosthenes measured the earth's circumference in the third century BC, finding it to be equal to 252,000 such stades. These units of the Hebrew order which measure the polar radius are allowed by the simple 10:63 ratio between the earth's least radius and its mean circumference to measure the latter also in significant integers.

In addition to being measured by the longer Greek and Roman units, the 97.320 96-ft. inner diameter of the Stonehenge circle is also equal to 28 sacred rods at the above value. The measured average width of the lintel stones is about $3\frac{1}{2}$ ft., allowing the supposition that they were intended to be one twenty-eighth part of the ring's inner diameter in width, or 1 sacred rod. In that case the outer diameter of the lintel ring is 30 sacred rods.

The three sets of dimensions to the Stonehenge sarsen ring, inner, outer and mean, were framed by the same units as measure the earth and relate proportionally to the earth's dimensions in terms of canonical numbers.

Polar radius   = 20 854 491 ft.
                     = 400 000 × Stonehenge outer radius
                     = 6 000 000 × Stonehenge lintel width
Mean circumference   = 131 383 296 ft.
                     = 2 520 000 × Stonehenge outer radius
                     = 2 700 000 × Stonehenge inner radius
                     = 37 800 000 × Stonehenge lintel width
                     = $12^4 \times 20$ × Stonehenge mean circumference
Mean radius   = 20 901 888 ft.
                     = $12^4 \times 20$ × Stonehenge mean radius

Summary of the earth's dimensions

Inherent in the units of traditional metrology is the code of geodesy which describes the slightly flattened sphere of earth in simple numbers and proportions. Written in decimals, the numbers seem long and arbitrary, so their fractional values are here added to show how they relate to the powers of 12 which constitute the framework of ancient mathematics.

polar radius	$= 20\,854\,491$ ft.	$= 3949.7142$ miles
	$= 12^4 \times 1000 \times 176/175$ ft.	$= 12^3 \times 2^4/7$ miles
mean radius	$= 20\,901\,888$ ft.	$= 3958.6909$ miles
	$= 12^6 \times 7$ ft.	$= 12^4 \times 21/110$ miles
mean circumference	$= 131\,383\,296$ ft.	$= 24\,883.2$ miles
	$= 12^6 \times 44$ ft.	$= 12^5/10$ miles

The mean radius is equal to $12^4 \times 1000$ shorter Greek feet of 1.008 ft. or to $6^3 \times 100\,000$ shorter Roman feet of 0.96768 ft.

No information on the length of the equator appears to be given by the above units of measure. It was no doubt in harmony with the other dimensions of the earth and related to them by whole-number proportions. Most likely is that it was taken as 1261:1260 with the meridian circumference. In that case the length of the equator would be 131 487 560, similar to the recently published figure of 131 484 632 ft. The likely figure for the equatorial radius is obtained by dividing half the equator by the π fraction, $22\,698/85^2$, thus making the length of the polar radius 20 926 902 ft. and its relation to the polar radius as 289:288.

Polar radius $\times\, 441/440$	$=$ mean radius	
$\times\, 289/288$	$=$ equatorial radius	
$\times\, 63/10$	$=$ mean circumference	
Mean circumference $\times\, 7/44$	$=$ mean radius	
$\times\, 1261/1260$	$=$ equatorial circumference	

Throughout the constructions of ancient sacred arithmetic there is interplay between two different values for π, 22/7 and 864/275. The figure given above for the earth's mean radius comes from dividing the mean circumference by 44/7. If 2π is represented by its other value, 1728/275, the length of the mean radius is given more simply as 3960 miles. The difference between that length and the alternative radius produced by 22/7 is 6912 ft. or exactly 6000 Egyptian feet of 1.152 ft.

Comparison of ancient and modern estimates of the earth's dimensions

	ancient metrology	*Encyclopaedia Britannica* 1983
polar radius	3949.7142 miles	3949.5 miles
mean radius	3958.691	3958.7
equatorial radius	3963.428	3963.5

The priority of the foot

The numbers which express the relationships between the various ancient units of measure, and those by which the units are multiplied to produce the dimensions of the earth, occur independent of any unit of reference because they are ratios. The values of the units themselves, however, conform to the number code of ancient science because they are in terms of the so-called English foot. With the foot as unit of reference researchers in ancient metrology are freed from having to rely entirely upon analysis of ancient instruments and monuments. The foot is the medium which shows up the numerical framework behind the values and ratios of the old units and allows them to be defined accurately in accordance with reason. Although it appears to stand apart from the other units, the foot is in fact the bond that unites them and makes them systematically coherent. It was the author's discovery some years ago (first set out in *Ancient Metrology*, 1981) that the English units are in geometric proportion with the old Greek and Roman units which made it possible to establish the exact values of the Greek and Roman feet, cubits, miles etc.

As calculated above, the longer Roman foot was equal to 0.973096 ft. and the longer Greek foot to 1.01376 ft. Five thousand of these units made up their respective miles, so the Roman mile was of 4866.048 ft. and the Greek mile of 5068.8 ft. The ratio between them is 24:25, and this ratio is also found to obtain between the Greek mile and the English mile of 5280 ft. The three mile units thus form a geometric progression:

Roman mile × 25/24 = Greek mile
Greek mile × 25/24 = English mile

It is so unlikely that this neat progression could have arisen by chance that the author feels justified in claiming it as strongly confirming the unit values previously arrived at by other considerations. By linking the English units with those of classical metrology it also justifies making the English foot the prime unit of reference. Modern researchers who accept the fiction that the English units are of recent origin, and adopt the habit of expressing the values of the old measures in terms of the new-fangled metre, thereby disguise from themselves the most significant aspect of ancient metrology, its basis in canonical number. Were it not for that aspect, the business of establishing the exact values of the old units would be of merely academic interest. The fact that those units, as here calculated in English feet, exhibit the same scale of numbers as found in ancient music and geometry is what makes the system of ancient metrology so relevant to the study of traditional science.

As seen above, the English units relate numerically to the earth's dimensions through the powers of 12, e.g. the $12^5/10$ miles in the meridian, but the most direct geodetic reference of the foot is to the equator. In archaic China and Babylon the circle of the equator was divided into $365\frac{1}{4}$ degrees to represent the number of days in a solar year (now taken as 365.2422 days). Each of the $365\frac{1}{4}$ degrees would measure some 360 000 ft., each minute 6000 ft. and each second 100 ft. If the number of days in the year is taken as 365.243 22, these figures become exact and accord with the estimate of the equator given above, 131 484 632 ft. Alternatively, dividing the equator into 360 degrees makes each degree equal to 365 243.22 ft. or the number of days in a thousand years.

Measured by the foot, the ancient units are found to be intervals or harmonies within a numerical frame, based upon duodecimal number with its nodes on the powers of 12, and also upon those numbers which describe the dimensions of the New Jerusalem circle, 504 and 3168. Taken in units of 10 000 to make them whole numbers, the shorter versions of the Roman and Greek units are multiples of 504. For example,

10 000 Roman feet	$= 96768$	$= 504 \times 192$ ft.
10 000 Roman cubits	$= 145 152$	$= 504 \times 288$ ft.
10 000 Greek feet	$= 100 800$	$= 504 \times 200$ ft.
10 000 Greek cubits	$= 151 200$	$= 504 \times 300$ ft.
10 Roman miles	$= 48 384$	$= 504 \times 96$ ft.
10 Greek miles	$= 50 400$	$= 504 \times 100$ ft.

The same units at their longer values in multiples of 10 000 000 are divisible by 3168.

10 000 000 Roman feet	$= 9 732 096$	$= 3168 \times 3072$ ft.
10 000 000 Roman cubits	$= 14 598 144$	$= 3168 \times 4608$ ft.
10 000 000 Greek feet	$= 10 137 600$	$= 3168 \times 3200$ ft.
10 000 000 Greek cubits	$= 15 206 400$	$= 3168 \times 4800$ ft.
1000 Roman miles	$= 4 866 048$	$= 3168 \times 1536$ ft.
1000 Greek miles	$= 5 068 800$	$= 3168 \times 1600$ ft.

The Egyptian and Hebrew units exhibit the multiples of 72 and the powers of 12, or are related to them by the 175/176 ratio. For example,

1000 sacred rods, shorter version	=	$3456 = 12^3 \times 2$ ft.
1000 Egyptian royal cubits	=	$1728 = 12^3$ ft.
10 000 Hebrew double feet, shorter version,	=	$20736 = 12^4$ ft.

The practical advantage of units of measure based on the numbers 5040 and 7920 (a quarter of 31 680) is that 5040 is 7!, the product of the numbers 1 to 7 multiplied together, while 7920 is 11!/7!, the product of the numbers 8 to 11. Thus the Greek and Roman units, being based on those numbers, had the maximum numbers of subdivisions.

Astronomy

The 360 degrees of a circle and the sexagesimal reckoning of time are relics of the ancient system of number in which all circular forms and cyclical movements were measured in units of 6 and 12. Each of the twelve gods in the circle of the zodiac ruled over a segment of thirty degrees, and as the sun at the Vernal Equinox entered a new sign, the energy of the world was said to be renewed. This happened every 2160 years, making the length of the Great Year, in which the sun completes its circuit through the zodiac, equal to 25 920 ($= 72 \times 360$) solar years.

The greater cycles of time, recognized in traditional Hindu chronology, are significant as marking periods of change in universal forms and psychology. Beginning with the second, of which there are 86 400 in a 24-hour day, periods of time build up to:

432 000 ($= 6^3 \times 2000$) years, the Kali Yuga
864 000 ($= 6^3 \times 4000$) years, the Dvâpara Yuga
1 296 000 ($= 6^4 \times 1000$) years, the Tretâ Yuga
1 728 000 ($= 12^3 \times 1000$) years, the Krita Yuga

One day in the life of Brahma is ten thousand times the length of the Kali Yuga, so 8 640 000 000 years is known as one day and night of Brahma.

Another relic of the archaic tradition that produced these divisions of time is our present system of measurement by units of feet, furlongs and miles, with the acre as the unit of land measuring. Those measures, which are still found the most convenient today, were canonized and held sacred, because not only do they relate both to the human and to the astronomical scale, expressing the unity between macrocosm and microcosm, but they bring out the same numbers in the dimensions of the solar system as were given to the units of time. The canonical dimensions of the earth, sun and moon are:

Diameter of sun	$= 864\,000$ miles $(6^3 \times 4000)$
Radius of sun	$= 432\,000$ miles $(6^3 \times 2000)$
Diameter of moon	$= 2160$ miles $(6^3 \times 10)$
Radius of moon	$= 1080$ miles $(6^2 \times 30)$
Mean diameter of earth	$= 7920$ miles $(6^2 \times 220)$
Mean circumference of earth	$= 24\,883.2$ miles $(12^5/10)$
Mean distance from earth to moon	$= 237\,600$ miles $(6^3 \times 1100)$
Mean distance from earth to sun	$= 93\,312\,000$ miles $(6^6 \times 2000)$

The dimensions and distances of the earth, moon and sun relate to each other by simple ratios in which the number 11 is combined with the duodecimal numbers of the Canon. Modern authorities differ slightly from each other in their estimates of the above astronomical distances, but their average figures are almost the same as those established by the ancients. The figures in parentheses below come from the *Encyclopaedia Britannica*.

	Earth diameter	Moon diameter	Sun diameter	distance Earth − Moon
Earth diameter 7920 miles (7917.4)				
Moon diameter 2160 miles (2160)	11 / 3			
Sun diameter 864000 miles (864950)	11 / 1200	1 / 400		
Earth − Moon distance 237600 miles (239900)	1 / 30	1 / 110	40 / 11	
Earth − Sun distance 93312000 miles (92957000)	11 / 129600	1 / 43200	1 / 108	11 / 4320

A striking illustration of the duodecimal framework of the solar system appears when the dimensions and distances of earth, sun and moon are calculated in feet (multiplying the mile figures by 5280). The diameter of the moon, the traditional measure of astronomical distances, is 2160 miles or 11 404 800 ft. The number 114 048, equal to 3168 × 36, is among the most important of the larger numbers in the numerical canon. The diameter in feet of the moon, 11 404 800, is multiplied by the better value of π, 864/275, to make the lunar circumference equal to 35 831 808 or 12^7 ft. Being in proportion with the moon, the other bodies and their orbits also feature 12^7 ft. in their circumferences. The Grand Orb referred to below is the circle made by the earth round the sun.

Taking $\pi = 864/275 = 3.14\overline{18}$,

circumference of	moon	$= 12^7$ ft.
	earth	$= 12^7 \times 11/3$ ft.
	sun	$= 12^7 \times 400$ ft.
	moon's orbit	$= 12^7 \times 220$ ft.
	Grand Orb	$= 12^7 \times 86 400$ ft.

The conclusion which emerges from these studies is that the ancient philosophers took the key 'New Jerusalem' numbers, 12 and 7, put them together in the form of 12^7, established that number in feet as the measure of the moon's circumference and made it the astronomical standard measure of the universe.

One of the beauties of their numerical system is the relative use they made of their two main versions of π, 22/7 and 864/275. In our earlier investigations of the New Jerusalem dimensions, and in the following account of Plato's Magnesia, can be seen how the alternate use of these two π fractions produces the appropriate sacred numbers in the groundplan of those cities. The same two fractions had an important function in astronomy, for (taking the moon's circumference of 12^7 ft. as the standard) they produce two significant sets of dimensions for the earth, sun and moon. The first, 22/7, gives to those three spheres their actual diameters, in accordance both with numerical harmony and with modern scientific estimates. The second, 864/275, provides convenient, whole-number approximations. The table below shows how the two versions previously given of the earth's diameter are reconciled through the two versions of π.

	circumference	*diam. ($\pi = 22/7$)*	*diam. ($\pi = 864/275$)*
moon	12^7 ft.	2159.288 95 miles	2160 miles
earth	$12^7 \times 11/3$ ft.	7917.318 miles	7920 miles
sun	$12^7 \times 400$ ft.	863 714.38 miles	864 000 miles

4 The cities of Plato

RUNNING THROUGH all Plato's works, most noticeably in the *Republic*, *Timaeus*, *Critias* and *Laws*, is the theme of an ideal city, alternatively presented as a cosmological scheme. In contrast to the honest, open, straightforward way in which he expounds his moral and philosophical reasonings, Plato's references to the cosmic city are always oblique and mysterious. He hints at knowledge of the code of numbers behind the structure of the universe, but it seems as if he was inhibited from imparting its details. According to his biographer, Olympiodorus, Plato studied sacred science with the priests in Egypt, where presumably he learned about the traditional canon of proportion referred to in the *Laws*, and he also went to Phoenicia to acquire the science of the Persian Magi. At home he was initiated into the Pythagorean system. It is likely, therefore, that one reason for his allusive treatment of the numerical science on which his doctrines were based was that the knowledge was not his own but had been entrusted to him under an oath of secrecy.

The 5040 citizens of Magnesia

The main subject of Plato's *Laws* is Magnesia, the city-state which he imagined himself founding in a depopulated area of Crete. He was never content with being a mere philosopher and pedagogue. Constantly in his mind was the problem of translating his knowledge into practical effect. Realizing how hopeless would be the task of persuading the mass of his fellow-citizens of the benefits to be gained by social reform on ideal principles, he concentrated his influence on those who were in a position to put his ideas into practice. At the court of Sicily he incurred the reputation of a dangerous eccentric by trying to convince the ruler, Dionysius, that he could make himself happy by devoting his life to the welfare of his subjects, and he wasted more time as tutor to the old tyrant's son in the hope of educating him as a philosopher. The courtiers intervened, and the ungrateful youth, hearing from them that Plato was scheming against him, not only dismissed his tutor but sold him off as a slave. Fortunately, a young Greek racing driver (of chariots) recognized a bargain and made himself famous as the man who purchased Plato and set him at large again. After that experience Plato must have realized how slim were his chances of finding a ruler qualified to play the part of philosopher-king.

Plato's political philosophy was radical-traditionalist, a combination of idealism and respect for custom. Its first principle was that state constitutions should be designed to imitate the cosmos: 'No state can find happiness unless the artist drawing it uses a divine pattern.' The pattern referred to is that which underlies every system in nature, from the order of the heavens to the constitution of the atom. It is a dynamic pattern, and to imitate it faithfully would require a reversion to nomadism with people wandering round their territories in regular orbits. Such a radical type of reform would have to be effected by violent means, as was attempted by Genghis Khan, who expressed his disapproval of civilization by razing cities and diverting river courses through their sites. Plato was more humane and more realistic. He recognized the advantages of settled life and set himself the task of reconciling that state with the human need to be in harmony with the seasons and patterns of nature. The ideal communist society, with no personal property and with everything held by the community in common, would not appeal to the respectable citizens whom Plato intended to be the first colonists of Magnesia. His city was therefore to be the 'second-best', a republic of equal small-holders and their families, each man provided with a decent education and a simple livelihood from an allotment of state land, which he was allowed to retain in his family and pass on to his heirs. In return he was obliged to undertake certain well-defined civic duties. The life of a Magnesian might have resembled that of Goldsmith's sturdy rustic in the happy days before 'sweet Auburn' became the Deserted Village.

> For him light labour spread her wholesome store,
> Just gave what life required, but gave no more:
> His best companions, innocence and health,
> And his best riches, ignorance of wealth.

Plato's approach to the founding of Magnesia was, first, to model its constitution on the perfect proportions of the canon of number as displayed in the New Jerusalem diagram. After that it was to be a matter of negotiation between the citizens and their 'lawgiver'. The Magnesians are told to encourage the founding philosopher to depict the ideal state, 'sparing no detail of absolute truth and beauty'. Then comes the time for compromising. 'If you find that one of these details is impossible in practice, you ought to put it on one side and not attempt it; you should see which of the remaining alternatives comes closest to it.'

In the latter part of the *Laws* Plato indulges his taste for practical constitution-making, defining the tendencies that should be encouraged and those to be punished among the citizens of Magnesia. Its code of laws and the duties of its officials are prescribed in great detail. But when he comes to describing the ideal pattern on which the state is to be founded, he is far less precise. The one thing he insists upon is that everything in the state, from the number of its citizens and land-holdings to the measurements of domestic utensils, must somehow be in accordance with a system of mathematics based on the number 5040. Several times Plato emphasizes the importance of founding all the state institutions on this number. First he points out the advantage of 5040 for the purpose of sharing out lands, goods and duties among the citizens, which is that 5040 can be divided by all the first ten numbers as well as the number twelve, and it has sixty divisors in all, including itself. Thus the 5040 land-holders of Magnesia can be divided and subdivided into the maximum number of equal groups, and the policy of fair shares for all is made as simple as it possibly can be.

As well as having a practical use in administration, 5040 serves as a symbol of that mystical code of number which Plato intended to be the guiding principle behind the foundation and government of Magnesia. Officials in charge of the state were to be initiated into the processes and philosophy of number, and it was to form the basis of children's education, thus raising the moral and mental qualities of each new generation of Magnesians. According to the *Laws* (747), 'no other branch of education has such a vast range of applications as mathematics; but its greatest advantage is that it wakes up the individual who is by nature sleepy and slow-witted, making him quick to learn, sharpening his memory and wits and leading him beyond his normal capacity by divine art'.

The foundation plan of Magnesia is given in *Laws* 745:

The lawgiver must plant his city in the centre of the country, choosing a spot which has all the other conveniences a city needs, which are easily learnt and specified. Next he must divide the land into twelve portions, after reserving a sacred area for Hestia, Zeus and Athena, calling it the Acropolis and building a circular wall round it. From there he must divide the city and the whole country into twelve sections. The division must be made equitable by arranging for those containing good land to be smaller than those with inferior land. He must mark off 5040 allotments, divide each into two parts and give every man a portion consisting of two pieces, so that each portion contains a near piece and a far piece – the piece nearest the city being put together with the one furthest away, the second nearest with the second furthest and so on with the rest. And these portions must be graded in the way previously mentioned, making them larger or smaller according to the quality of the land, so that everyone has a fair share. The lawgiver must also divide the citizens into twelve sections and share out equally among them all the rest of the property other than land, according to its value after that has been assessed. Finally, the twelve land divisions are to be dedicated to the twelve gods, consecrating each section to each god as decided by lot and calling each a tribe. And the city must also be divided into twelve sections, as was done with the rest of the land, so that each citizen is allotted two houses, one near the centre and the other near the periphery. That will complete the foundation of the colony.

In each of the sections a proportion of the land is set aside for administrative buildings, villages, markets, temples, shrines and monuments. 'The temples we must erect all round the market-place and in a circle round the whole city, on the highest spots . . .' (*Laws* 778). There is to be no actual city wall, because that would make the citizens less ready and watchful, but a continuous ring of private houses round the perimeter of the city will provide a barrier against the outside world. 'The whole city', says Plato mysteriously, 'will have the form of a single house.'

The general form of Magnesia is evidently a large circle, and within it is the smaller ring of the city surrounding the acropolis at the centre. Its whole area is divided into twelve unequal segments by lines radiating from beyond the walls of the central acropolis. Their inequality is a matter to be considered later. For the present, in order to establish the principle behind the division, we divide the circle into twelve regular segments (excluding the acropolis), each segment containing a twelfth part of the city and a twelfth part of the land beyond. The whole area of the land is then cut up into 840 parts by 840 equally spaced rings drawn from the centre, and the city is similarly divided. In that way each of the 420 citizens (a twelfth of 5040) of the tribe occupying one of the segments receives four bits of land. To the first is given that portion of the outer ring which lies within his tribal territory, together with the corresponding portion of the ring outside the city. Within the city he also has the outer portion and the inner portion. The next member of the tribe has the

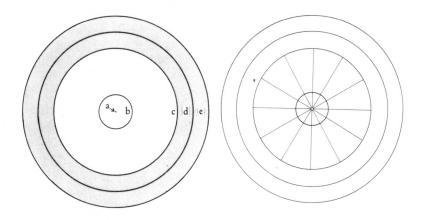

Figure 45. The five rings of Magnesia: acropolis, city, cultivated land, surrounded by two rings of hills and temples.

two portions of the land and the two portions of the city adjacent to those of the first, and so on. The arrangement is made clearer by the diagram (figure 46).

In common with the New Jerusalem, Magnesia is a structure of concentric circles, and Plato's emphasis on the number 5040 indicates that both cities are versions of the same archetypal plan. It has previously been shown (pages 44–5) that the different areas in the New Jerusalem diagram, measured by the so-called NJ unit (3168/7 sq. ft. or the area of a circle with radius 12 ft.), represent multiples of 504 or 5040 NJ units. These areas are therefore perfectly adapted to accommodate the 5040 citizens of Magnesia. Given below are the dimensions of the principal rings comprising Magnesia, together with their symbolism compared to that of New Jerusalem. Areas are given in NJ units.

Radius of circle *Area of circle* *Width of ring* *Area of ring*

	Radius of circle	Area of circle	Width of ring	Area of ring	
(a)	72 ft.	36	72 ft.	36	
(b)	1 080 ft.	8 100	1 008 ft.	8 064	= 5040 × 16
(c)	3 960 ft.	108 900	2 880 ft.	100 800	= 5040 × 20
(d)	5 040 ft.	176 400	1 080 ft.	67 500	
(e)	6 120 ft.	260 100	1 080 ft.	83 700	} 151 200 = 5040 × 30

Magnesia
(a) temple acropolis
(b) city
(c) cultivated land
(d,e) hills with temples

New Jerusalem
(a) universal axis
(b) the underworld
(c) the earth
(d,e) sphere of the moon

The plan of Magnesia which satisfies Plato's main conditions, apart from that of making the twelve segments irregular, is shown in figure 45a. All the land is available for the citizens except the inner acropolis and the double outer ring. In the New Jerusalem these outer rings enclose the twelve lunar circles and represent the heavens. In Magnesia they are the sacred heights surrounding the city where, says Plato, a ring of temples is to stand. This area should not, therefore, be used as allotments.

The land which remains available for distribution consists of the ring which has a width in cross-section of 2880 ft., designated above as cultivated land, together with the ring of the city, radius 1080 ft., from which is deducted the inner acropolis, radius 72 ft., making the width of the city ring 1008 ft. These combined rings represent an area of 108 864 NJ units, so each of the 5040 members of the Magnesian community has a total area of 21.60 NJ units, a hundredth part of the familiar sexagesimal number 2160.

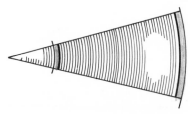

Figure 46. One of the twelve segments of apportioned land in Magnesia showing the principle of division. 840 arcs on the cultivated land and the same number on the city allow each of the 420 citizens of a tribe four strips of land, two in the country and two in the city, the largest strip being matched with the smallest so that the division is equal. A specimen holding in the country is marked in black in the diagram.

According to Plato each citizen has four holdings, two in the city and two in the surrounding country. For this purpose the land is divided into 840 rings, and so also is the city up to the acropolis. Lines drawn outward from the acropolis cut the whole area into twelve segments. In the city and the country each citizen has a double holding consisting of:

cultivated land, 100 800 NJ units; divided by 5040 = 20   NJ units
city and market,   8 064 NJ units; divided by 5040 = 1.6 NJ units
total area   108 864 NJ units; divided by 5040 = 21.6 NJ units

Now can be seen clearly the function of the central acropolis, of area 36 NJ units, in the numerical constitution of Magnesia. If the area of the whole circle, radius 3960 ft., is calculated with the use of 864/275 as π, it amounts to 108 864 NJ units. In that case the acropolis circle is included in the area. If, however, the more convenient $\pi = 22/7$ is used, the whole area contains 108 900 NJ units. The number 108 900 is not divisible by 5040, so the acropolis circle is omitted, and the result is $108\,900 - 36 = 108\,864$ NJ units, allowing 21.60 units to each of the 5040 Magnesians.

Yet this attractive arrangement can not be claimed as the perfect reconstruction of Magnesia because it ignores two of Plato's conditions. The land, he says, is not to be divided into twelve exactly equal segments, but 'the division must be made equitable by arranging for those (segments) containing good land to be smaller than those with inferior land'. Plato also

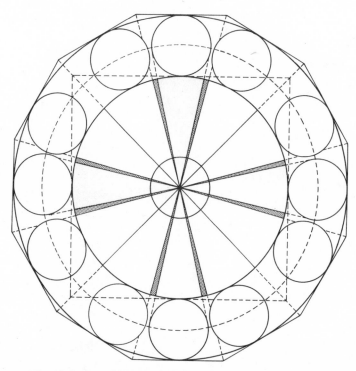

Figure 47. The ideal plan of Magnesia takes the form of a wheel cross over the New Jerusalem diagram. The strips bordering the arms of the cross are taken from the land they enclose, making each of the enclosed four segments less by one-fifth than the neighbouring segments.

refers to thin strips of no-man's-land which in some cases separate the twelve segments occupied by the twelve Magnesian tribes.

Contemplation of the New Jerusalem diagram, the ancient geometers' image of the cosmos which is the source of Plato's Magnesia, brings light to these problems. In figure 47 the whole city is divided into twelve equal parts by twelve radii, making the angle at the centre between two adjacent radii 30°. Eight further lines are then drawn from the centre. They are arranged in four pairs, each pair enclosing one of the four lunar circles which are positioned north, south, east and west within the outer perimeter of the diagram. Calculations prove that the angle made by each pair of these lines is slightly over, but for the purposes of geometry close enough to, 24°. Between these eight lines and eight of those which make the twelve-part division of the circle, eight narrow segments are left over. The angle which they each form at the centre is about 3°. The circle which contains the city and the cultivated land of Magnesia is thus divided into 20 unequal segments, consisting of

8 segments of 9072 = 72 576 NJ units
4 segments of 7257.6 = 29 030.4 NJ units
8 segments of 907.2 = 7 257.6 NJ units
total area   = 108 864 NJ units

The eight narrow segments answer Plato's requirement that an area of no-man's-land be in some cases inserted between the boundaries to two districts, and that the allotments of good land should be smaller than those made up of the poorer ground. It seems that Plato calls the good land those segments lying north, south, east and west on the diagram, while the inferior districts are those which are placed at an angle to the cardinal axes. Some unknown theory of augury may lie behind this characterization. Whatever the reason, the four segments forming a cross over the diagram are smaller by a fifth than the eight others, sacrificing some of their land to the strips of common ground bordering them, which are perhaps meant to represent roads, ditches and other communal necessities. They are generally more accessible to those people in the smaller districts than to those who share a boundary with their neighbours. The principle behind the unequal division may therefore be that those who live more remote from access to the centre should be compensated by a larger holding.

The final division of the cultivated land and the city between the 5040 citizens of Magnesia works out as follows:

In the smaller sections each has in the country 16 NJ units
in the city 1.28 NJ units
In the larger sections each has in the country 20 NJ units
in the city 1.6 NJ units

Thus the 1680 citizens of the 4 tribes occupying the good land each have a cultivated plot of 16 units, while the 3360 members of the other 8 tribes each have 20. When the land in the city is included the total areas of the allotments are:

smaller portions each of 17.28 = 100th part of $12 \times 12 \times 12$ units
larger portions each of 21.60 = 100th part of $6 \times 6 \times 60$ units

From circle to square: the City brought to earth

Just and aesthetic though it is, the arrangement of Magnesia in hundreds of concentric rings is obviously most impractical and would present a daunting task to its surveyors. Practical land division is square or rectangular. The City as a circle represents the heavenly ideal or archetype. That is the model which the rulers of the state are required to study, thus refining their mentalities and imbuing them with the sense of proportion which enables them to govern justly. However, when it comes to planning and running an actual city, Plato tells them to compromise by reducing the ideal to the practical. The heavenly circle is to become the earthly, rationally measured square.

As previously shown, the populated area of Magnesia in its ideal form, containing the cultivated land, city and acropolis, lies within a circle of radius 3960 ft. Around it is the high ground with temples. The circle is enclosed by a square of 1440 acres, and that is the shape of practical Magnesia.

The problem which now arises is how to divide up the 1440 acres of the square in the same proportions as in the circular city. It is easy enough to work out the area occupied by each of the two categories among the tribes and by the common lands. Four of the tribes each have four parts of land to the five parts held by each of the other eight tribes, so the division is:

8 tribes each with 120 acres = 960 acres
4 tribes each with 96 acres = 384 acres
8 strips of common land, each of 12 acres = 96 acres
Total = 1440 acres

To each of the 8 tribes is thus given a hide of 120 acres, the same as was allotted to each of the twelve founding saints at Glastonbury. The four tribes have a smaller hide of 96 acres and the remaining 96 acres are for communal

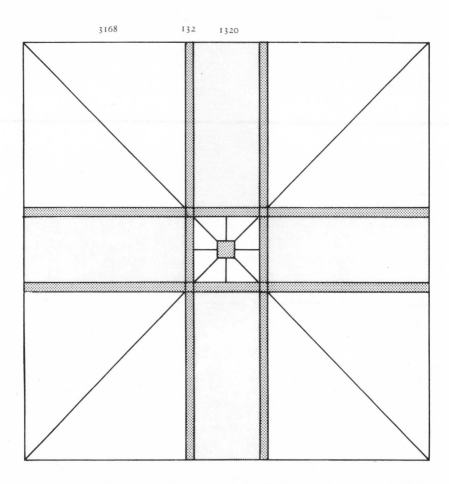

3168   132   1320

Figure 48. In the plan of Magnesia as a square the land is divided in the same proportion as in the circular city. The four tribes in the favoured positions within the four arms of the cross each occupy a rectangular area of 96 acres, and the eight other tribes each have two pieces of land (A,a) amounting to 120 acres. Four strips of common land contain altogether 96 acres, the squares where they overlap being placed together at the centre.

This diagram also fits the description of the Holy City given in the last chapters of Ezekiel.

use. For practical purposes these areas are obviously insufficient. However frugal the citizens of Magnesia, all 5040 of them could hardly inhabit
and live off a mere 1440 acres. Plato is apparently describing the microcosm of a city, a ritual amphitheatre for assembling the twelve tribes of
Magnesia.

The advantage of its number system and the way in which its lands are
proportioned is that Magnesia can be represented in many different ways as a
symmetrical plan. For a number of reasons Figure 48 is proposed as the
image of Plato's ideal city. As a square version of circular Magnesia it
reproduces the proportions of the original.

In drawing it, the 12-furlong sides of the 1440-acre square are each divided
into five equal parts, and the middle section forms the cross over the figure.
Within its borders on either side are the four strips of common land, and the
four squares made where they overlap are placed together at the centre. The
width of these strips is 132 ft. or two chains, and their total area is therefore
96 acres. The four rectangular areas within the arms of the cross, north,
south, east and west, measuring 2 by 4.8 furlongs (1320 × 3168 ft.), contain
96 acres each and so represent the lands belonging to the four tribes. The
eight tribes are lodged in pairs, each pair occupying one of the corner squares
together with the corresponding area within the square formed by the
crossings of the strips of common land. Thus each of the eight tribes has
115.2 + 4.8 = 120 acres.

The only feature of the circular city not yet accounted for in its square
version is the central acropolis. When the lands of Magnesia were divided in
segments, a proportion of the acropolis had to be deducted from each
segment, which had the effect of making each separate area of the city
divisible by 5040 NJ units. In the square city the area corresponding to the
acropolis is placed at the centre within the block of common land, so there is
no need to consider it when calculating the areas of the tribal lands. These
lands, consisting of either 120 or 96 acres (11 550 or 9240 NJ units), are so
divided that every one of the 420 citizens in each of the larger territories
receives $27\frac{1}{2}$ NJ units and those in the smaller territories receive 22 units.

Plato's secrecy

The number code running through these figures, implied by the number
5040, was certainly in Plato's mind when he composed Magnesia, and
presumably the diagrams in front of him when he described it were very

similar to those shown here. The question of why he merely hinted at his knowledge of the ancient science, elaborating the philosophy he derived from it without ever openly revealing its contents, leads to the very heart of his thoughts and motives. No doubt he was initiated into the science of proportion by certain teachers on the condition that he should not lightly disclose it; but he later freed himself from the Pythagorean cult of secrecy, and he certainly did not lack the courage and honesty to make public whatever he considered to be of public benefit. The inner core of his philosophy would have been discussed among his private pupils, those whom he could trust not to discredit the knowledge imparted to them, but when Plato tried for political influence by instructing at the court of Dionysius, he was sadly disillusioned. On the strength of what he had picked up from other teachers, the tiresome young tyrant fancied himself an accomplished philosopher, with no need of further assistance. He even presumed, much to Plato's disgust, to write his own philosophical treatise. Through this and doubtless other experiences with pretentious idiots Plato was unwilling to publish the mathematical gist of his teaching. In his Seventh Epistle he gave reasons for this decision.

> There does not and never will exist any treatise of mine dealing therewith [with the subject of Plato's serious studies]. For it is incapable of being expressed in words, unlike other studies, coming to birth in the soul as the result of constant application to it and communion with it, and it comes suddenly, as light is kindled by a leaping spark, and thereafter it nourishes itself.

If he could have written openly, he said, he would have done so. But the result would have been to arouse the contempt and enmity of pedants and to encourage in silly people 'vain aspirations as if they had learnt some sublime mystery'.

'And this is the reason', said Plato, 'why every serious man in dealing with serious subjects carefully avoids writing about them, lest he may thereby cast them as a prey to the envy and stupidity of the public' – as pearls before swine.

Since we are here attempting and purporting to elucidate some of Plato's mysteries, his words are somewhat disconcerting. In his day, however, there were others in the world who possessed the knowledge which Plato concealed, and he must have hoped that they or their successors some time in the future would be able to use that knowledge in the way he himself sought to use it, for the reform of education and the restoration of that traditional cosmic standard which controlled and sanctified the ancient world-order. If they exist today, if there are people or groups who have inherited the secrets of the ancient canon, their influence is not very apparent. Clearly apparent, however, is the modern need for those very qualities which the canon of proportion was supposed to impart to societies which adopted it, qualities of endurance, equilibrium and harmony under natural law. The source of those benefits is thus a legitimate and timely object of study. At the very least it is harmless – which is more than can be said for many of the theories and systems which are now peddled in the market and flourish in high places. As further justification of these studies one can quote Plato himself, saying that 'a man who knows of a subject which he finds sublime, true, beneficial to society and perfectly acceptable to God, simply can not refrain from calling attention to it.'

To conclude the subject of Magnesia, figures 47 and 48 are offered as Plato's two versions of its groundplan, the first being its ideal, unmanifest circular form, the second its earthly, actual shape as a square. Common to both forms is the one code of number which is the only invariable element in the city. However one measures its areas, whether by our so-called New Jerusalem unit or by any of its parts or multiples, the same series of numbers determines all its dimensions and divisions. Within the framework of its definitive numbers Magnesia can be constructed as any regular shape and on any scale, but however it is actually built, whether as a single house, a village, a city, a state or a new world-order, the archetype it refers to is heaven and earth, symbolized by the perfect proportions of the New Jerusalem plan.

On the astronomical scale, Magnesia and the New Jerusalem represent the sublunary world, that is, the earth and its atmosphere under the influence of the moon. In two of his other geometrical allegories Plato went further and described the whole universe. One of these, the extraordinary tale of a man coming back from among the dead called the Myth of Er in the *Republic*, is based on the New Jerusalem diagram and develops from it. The other, in *Timaeus*, is about the composition of the World-soul. In this Plato uses the same numbers as in his other schemes, but applies them to music rather than measurement.

In contrast to these ideal images, Plato's Atlantis, described in *Critias*, is subtly incorrect, a close imitation but a travesty of the ideal, constructed from sacred numbers misapplied. The mathematical reason for the fall of Atlantis is given later in this chapter.

Atlantis

Atlantis was a magnificent city, but it did not quite work and finally it came to grief. The story of its destruction is briefly told in the early part of *Timaeus*, and in *Critias* Plato describes its foundation and how it was designed. But unfortunately, just when he reaches the point in the tale where the Atlanteans are about to be punished by the gods for their transgressions, he breaks off, and the book is left unfinished. Modern believers in Atlantis and cosmic catastrophes, who have flourished since 1882 when Ignatius Donnelly's classic *Atlantis: the Antediluvian World* was first published, are particularly disappointed by this. Plato was certainly one of their school, holding the traditional belief that civilization is periodically destroyed by cataclysms due to some natural, or divinely willed, event such as the approach to earth of an erratic heavenly body. In the last sentence of *Critias* he tells how Zeus called together all the other gods to judge the fate of Atlantis, 'and when he had assembled them, he spake thus: . . . '

The speech that followed being unrecorded, there is no telling by what mechanism the gods produced the upheavals that devastated the Mediterra- nean countries and caused the island of Atlantis 'to be swallowed up by the sea and vanish'. The reason why the Atlanteans incurred such harsh treatment is, however, stated.

> When the element of divine nature within them became weak and diluted through being constantly mixed with large portions of mortal nature, so that the human element finally became dominant, then they lost their fair qualities through being unable to bear the burden of their riches and began to appear ugly to those who have the gift of perception.

As a contrast to Atlantis, Plato in *Critias* first describes the ancient state of Athens, which preserved the nobility of its character up to the time when its army was engulfed by the Atlantis flood. The catastrophe happened about 9000 years before Plato's time, following a great battle in which the Athenians had defeated the Atlantean hosts, thwarting their ambition to conquer the entire Mediterranean area. The constitution of ancient Athens resembled those of Plato's ideal cities in the *Republic* and *Laws*. It was founded by two gods, Hephaistos and Athena (whose combined numbers, 795 + 69, amount to the 'foundation number', 864), and its 20000 citizens lived in simple, orderly fashion, strictly in accordance with the perfect code of law left to them by their divine progenitors.

Atlantis was also divinely founded, being under the tutelage of Poseidon; but the god had taken a native woman as his wife. The rulers of Atlantis were descended from that union, and from the very beginning their stock was tainted with the fallible, mortal element which eventually predominated in them. Poseidon and his wife, Kleito, had ten sons, five pairs of twins. Each was given a portion of the island for him and his descendants to administer, the eldest taking the home territory and supremacy over the others. They ruled as a confederacy, meeting in the Temple of Poseidon every fifth and sixth year alternately (in order, so Plato says, to give equal honour to odd and even numbers), where they rehearsed the laws of their founder, engaged in a ritual bull-hunt and sacrifice and spent a night of vigil. Atlantis became exceedingly prosperous, both through its own products and the profits of trade, but for many generations the kings maintained the traditional style of rule, the Atlanteans were uncorrupted by their wealth and they lived in the same virtuous manner as the ancient Athenians.

The island of Atlantis was of great size, bigger than Libya and Asia Minor put together. It lay beyond the Pillars of Hercules, and a series of other islands beyond it made it easy for a traveller to cross the ocean to the continent on the far side, the natural meaning of which is America. It was highly fertile and rich in animals, including elephants, in plants and in minerals. Near the middle of its southern side, almost down to the coast, stretched a great plain, 3000 stades long on its north and south sides and 2000 stades long on its sides facing east and west. Over many generations the Atlanteans had shaped the plain into a regular rectangle by digging a canal round it, 100 feet deep and one stade wide. The entire enclosed area was subdivided by channels 100 feet wide which ran straight across the plain from north to south at intervals of 100 stades from each other. They were intersected by cross-channels east to west, and the plain was divided into 60 000 citizens' allotments each of 100 square stades. On three sides the plain was surrounded by mountains, 'more numerous, lofty and beautiful than any which exist today', full of timber, domestic animals and game, which supported a large population in prosperous villages.

On the seaward side of the plain, near its mid-point, was the low hill where Kleito's parents had lived. Her union with Poseidon was formed soon after their death, and the couple made their home on her inheritance, where their ten twin sons were born. Posedion fortified their dwelling by making it an island, surrounded by three equidistant rings of water and two of land. At some later period the Atlanteans changed and added to this work. In the middle of the central island they built a magnificent palace around the original shrine and marriage-bed of Kleito and Poseidon, and made a road to and from it. They next dug a canal 300 feet wide from the outer ring of water to the ocean 50 stades distant, and continued it inwards up to the island by cutting through the intervening rings of land, following the same line as the roadway which therefore ran directly above it. The dimensions of the city's rings of land and water are given by Plato as follows:

central island, diameter 5 stades
inner ring of water, width 1 stade
ring of land, width 2 stades
second ring of water, width 2 stades
ring of land, width 3 stades
outer ring of water, width 3 stades

The total radius from the centre of the island to the outer rim of the far ring of water was therefore $13\frac{1}{2}$ stades, and the distance from the centre to the sea $63\frac{1}{2}$ stades. From the point where the canal met the sea a circular wall was constructed, running right round the city at a distance of 50 stades from its outer rim. Its radius was therefore $63\frac{1}{2}$ and its diameter 127 stades.

Atlantean architecture was of great splendour. The houses and buildings of the city were of different coloured stones, red, white and black, laid in ornamental patterns. Walls surrounded each of the rings of land, the outer one covered in brass, the next in tin and the wall surrounding the island acropolis in orichalcum, 'mountain copper', the sparkling metal peculiar to Atlantis. The central shrine was encompassed by a wall of gold. No doubt these metals had planetary significance, and there is a further hint of astronomical symbolism in the circular racecourse, one stade wide, which went round the middle of the outer ring of land as if in imitation of Phoebus's daily circuit of the heavens.

On the outer rings of land were many temples, gardens and places for recreation and exercise, and the island acropolis was made delightful with lakes and groves nourished by hot and cold springs which also provided baths. There stood Poseidon's gaudy temple, 300 feet wide by 1 stade in length, coated in silver, ornamented by many precious metals and with pinnacles of gold. Statues of 100 nereids on dolphins surrounded the gigantic image of the god within the temple, and around it on the outside were golden images of all the descendants of the original ten kings, together

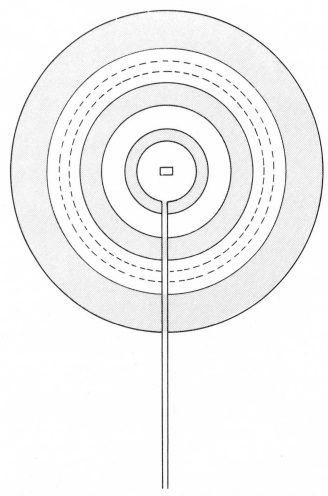

Figure 49. The city of Atlantis with its rings of water (shaded) and land. A racecourse runs round the outer land ring, and the Temple of Poseidon is at the centre of the inner island. The canal to the sea is bridged on its course through the rings, providing a road to the centre.

with their wives, and other sacred monuments. Yet for all the magnificence of the temple, says Plato, 'there was something barbaric about it.'

The inner island also contained the founders' shrine, the marriage-bed of Kleito and Poseidon, 'set apart as holy ground and encircled with a wall of gold, this being the very spot where at the beginning they had generated and brought to birth the family of the ten royal lines'.

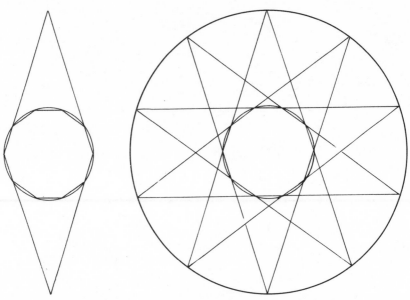

Figure 50. The birth of Kleito's first pair of twins is geometrically represented by two horns generated from a decagon, one side of which is taken as measuring 60 ft. With the other four sets of twins the ten-pointed figure is complete, and it is enclosed in a circle with a radius of virtually 300 ft. The diameter of this central shrine is therefore 600 ft. or 1 stade, which is the length of the Temple of Poseidon. Thus the Temple was in proportion with the circular shrine.

This last sentence, together with the decimal symbolism of the temple and other Platonic hints, make it plain that the foundation plan of Atlantis was based on the geometry of the number ten. It was generated from the marriage-bed which must therefore have stood at the centre of all, probably enclosed by the temple. The geometric image of the five pairs of twin sons or ten royal lines is a ten-pointed star constructed from ten straight lines, with five pairs of lines representing the ten twins. Their geometrical generation is shown in Figure 50. From their circular marriage-bed the founding couple gave birth to a decagram, springing from a decagon within the original circle and forming a greater circle. The following enquiry into the scheme of proportion linking the various parts of Atlantis points to the conclusion that the circle containing the ten royal lines had a diameter equal to the length of the temple, 1 stade or 600 feet.

Two further diagrams complete the development of Atlantis according to the geometry of the number ten, each stage representing a new royal

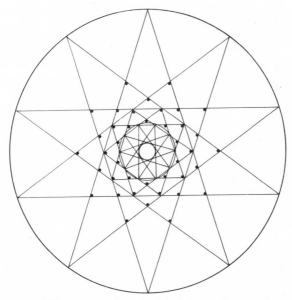

Figure 51. The ten-fold geometry of the central shrine, radius 300 ft., expands to the circumference of the central island of Atlantis which has a radius of 1500 ft. or 2½ stades.

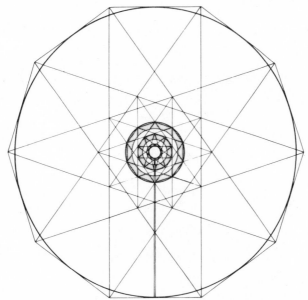

Figure 52. The Atlantean ten-fold geometry develops beyond the central island and the ringed city, forming a decagon which encloses the outer wall of the settlement.

generation. First (figure 51), the pattern expands so as to fill the central island. At a later stage it produces a decagon within the outer ring of Atlantis city, and finally (figure 52) it reaches the circular wall round the whole settlement, enclosing it within a greater decagon.

It is a splendid looking scheme, but as Plato said there is something barbaric about it. It lacks classical simplicity, being elaborate and difficult to construct accurately, which is the disadvantage of ten-fold as compared with simple twelve-fold geometry. Already we begin to see the cause of Atlantis's downfall.

The measures of Atlantis

The three geometrically important circles in the Atlantis scheme are, first, the inner circle of the ten twins enclosing the sacred marriage-bed, next, the circle enclosing the city, and finally the outer ring wall. The corresponding circles on the Magnesia diagram are the temple acropolis, the city and the circle of cultivated land. It is clear from Plato's account that the Atlantean unit of measure was the stade of 600 ft. rather than the 660 ft. furlong of

Figure 53. Ten-fold geometry as used in Atlantis. Its characteristic angles are of 36°, 72°, 108°, 144°.

Magnesia. In the land measurements of Atlantis the same units appear as in Plato's other diagrams, the acre of 43 560 square ft. and the related NJ unit. That unit best illustrates the relative proportions of the parts of the city.

The list below gives the dimensions of each of the individual concentric rings in Atlantis, including those of land or water which comprise the city itself. The first column repeats the width in stades of each ring, as listed previously, the second gives the extreme distances from the centre, and in the third column are the calculated areas of the separate rings in NJ units. It is notable that each area is a multiple of a basic unit, which is the area of the inner circle of the ten royal lineages, 625 NJ units. This number is appropriate to Atlantis which was designed by the numbers 10 and 5, because $625 = 5^4$.

	width (stades)	distance from centre (stades)	area (NJ units)
inner circle	$\frac{1}{2}$	$\frac{1}{2}$	625
island	2	$2\frac{1}{2}$	$15\,000 = 625 \times 24$
water ring	1	$3\frac{1}{2}$	$15\,000 = 625 \times 24$
land ring	2	$5\frac{1}{2}$	$45\,000 = 625 \times 72$
water ring	2	$7\frac{1}{2}$	$65\,000 = 625 \times 104$
land ring	3	$10\frac{1}{2}$	$135\,000 = 625 \times 216$
water ring	3	$13\frac{1}{2}$	$180\,000 = 625 \times 288$
outer wall	50	$63\frac{1}{2}$	$9\,625\,000 = 625 \times 15\,400$

That last figure which represents the entire area of Atlantis within the outer wall, omitting the area of the ringed city, is of remarkable significance. The 9 625 000 NJ units are equivalent to 4 356 000 000 square ft., which is a familiar number in metrology because 43 560 square ft. make up one acre. The total area of the enclosed land in Atlantis beyond the city limits was therefore exactly 100 000 acres.

Further interesting results are obtained by measuring the entire area of Atlantis within its wall, radius 63.5 stades, including the city but excluding the central shrine, radius 300 ft. The figures are:

	sq. stades	sq. feet	NJ units
area within wall	12 672.7857	4 562 202 857.143	10 080 625
central shrine	.7857	282 857.143	625
difference	12 672	4 561 920 000	10 080 000

The significance of these figures is that they are clearly designed to feature

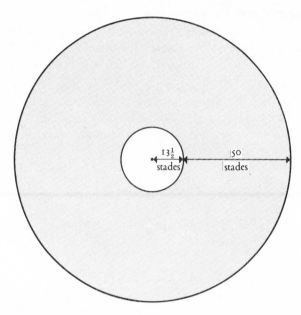

Figure 54. Deducting the area covered by the ringed city (inner circle), the amount of land contained by the outer wall of Atlantis is 100 000 acres exactly. (1 stade = 600 ft.; 1 acre = 43 560 sq. ft.)

those canonical numbers which are the principal elements in the numerical composition of Magnesia and New Jerusalem, the numbers 3168, 5040 and 1440; for the area of the whole of Atlantis within its outer wall, deducting the area of the founders' sacred shrine at the centre, is:

in square stades 12 672 $= 3168 \times 4$
in square feet, 4 561 920 000 $= 3168 \times 1 440 000$
in NJ units, 10 080 000 $= 5040 \times 2000$

Doomed Atlantis was founded on the very same numbers as those which govern the dimensions of its antithesis, the exemplary city of Magnesia. If numbers in themselves had any special magic, one would expect Atlantis to have been as nearly immortal as Magnesia was supposed to be. The fact of its eventual destruction proves that there are no numbers with miraculous powers, and that the use of canonical numbers such as 5040 and 3168 brings no automatic benefit. Atlantis was modelled on the same archetype as Magnesia, and the same canon of ideal cosmology provided the numbers and proportions in both cities. The foundation pattern of Atlantis, however, reflected a less beneficial aspect of that archetype than did Magnesia. That means that from the very beginning it was flawed. The matter is discussed later, concluding with the mathematical reason for the fall of Atlantis.

The problem of the plain

A certain feature which is essential to a foundation scheme of sacred geometry has not yet been discovered in the design of Atlantis. The city is all in rings and has no square figures in it. It appears therefore to lack the necessary squared circle. But it stood on a plain which was all divided into squares. There is no obvious geometric link between the circular city and the rectilinear plain. Their relationship remains to be established, and the solution of that problem reveals the missing squared circle figure.

The plain surrounding Atlantis was a natural rectangle measuring 2000 stades north to south and 3000 stades east to west. That shape had been made perfectly regular by enclosing it in a rectangular frame formed by a canal 1 stade wide and 10,000 stades long, or 2000 stades by 3000 stades. From the canal bordering the northern edge of the plain smaller channels, each 100 feet wide, spaced at intervals of 100 stades, ran due south across the plain and into the canal along its southern side. The pattern over the plain was evidently a grid, for Plato says that the Atlanteans 'cut cross-channels between them (the north-south channels) and also to the city'. He does not say whether the cross-channels were of the same width and spaced at the same intervals as the others, but if they were, 29 channels would run north to south and 19 east to west, dividing the plain up into 600 equal squares. That seems to have been the case, for according to Plato the plain contained 60,000 allotments, each of 10 times 10 stades.

At first sight the arrangement looks simple. The total area enclosed by the canal frame is 6 000 000 square stades and the number of holdings 60 000, leaving each citizen with 100 square stades and each square island with 100 such, placed in a block of 10 by 10.

But it is not in fact as easy as that, and a contractor who took on the task of digging the canals on the Atlantis plain to these specifications would soon be in trouble. In the first place, no provision has been made for the amount of land taken up by the channels. There is also the matter of the area occupied by metropolitan Atlantis, the ringed city. When these areas are deducted from the amount of land available for distribution, it is no longer possible to

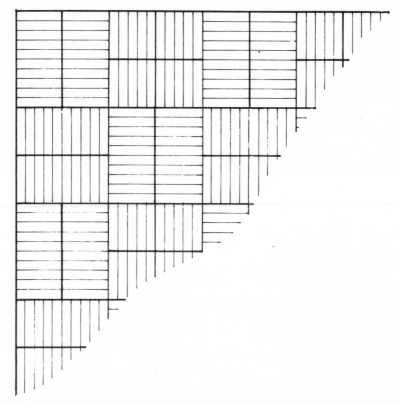

Figure 55. A means of dividing up the Atlantean plain so that each of the 100 citizens occupying one of the square islands would have an equal share in the land and water areas.

allot 100 square stades to each of 60 000 citizens. However one contrives it — by apportioning parts of the water-channels to each holding, by proportional reduction of the squares or by any other ingenuity — Plato's mathematical problem seems beyond any possible solution. Yet it was not his habit to waste breath on meaningless facts and figures. In all his geometrical allegories the stories are accompaniments to the construction of diagrams. In the case of Atlantis within the wall, the conformity of its development to the construction of a decimal scheme of geometry has already been shown, and Plato's carefully worded statements about the disposition of the plain, though on the surface paradoxical, were surely not intended as nonsense but to provide teasing clues to the full development of Atlantean geometry.

The problem of the plain having no direct solution it must be approached

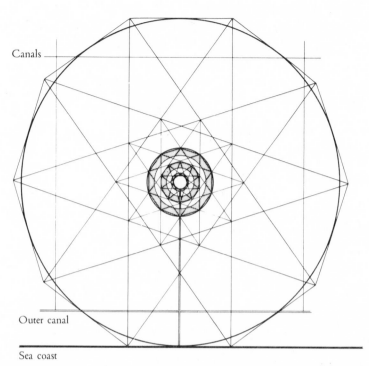

Canals

Outer canal

Sea coast

Figure 56. Placed in one of the squares on the plain formed by the lattice of water channels, Atlantis exhibits the squared circle, but it is not quite perfect, because the perimeter of the square made by the channels is 400 stades whereas the circumference of the outer wall is only 399½ stades.

from an angle, and the key to it lies in considering the position of Atlantis in relation to its surroundings. In *Critias*, 118, it is said that the country around Atlantis 'rose sheer out of the sea to a great height, but the part about the city was a flat, even plain, enclosing the city and being itself encircled by mountains which stretched as far as the sea'. This clearly means that the city stood within the borders of the plain and not, as Platonic editors sometimes depict it, beyond them. It was built on a low hill, 'near the middle of the plain and about fifty stades inland'. Fifty stades is the distance from the sea to the outer, water-filled ring of the city, so the middle of the plain must mean the middle of its seaward edge. But Plato's words are that it was *near* the middle rather than precisely central. That implies that Atlantis was placed to one side of the plain's north-south axis. It stood therefore in the very centre of one of the island squares bordered by waterways, as shown in figure 56.

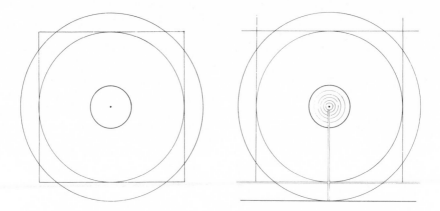

Figure 57. The squared circles of Magnesia (or the New Jerusalem) and Atlantis compared. In each case the difference between the radii of the two outer circles is equal to the radius of the inner circle. Yet the Atlantean squared circle is slightly defective.

Magnesia:		*Atlantis:*	
inner city, radius	= 1080 ft.	city, radius	= $13\frac{1}{2}$ stades
circle within square,		circle within square,	
radius	= 3960 ft.	radius	= 50 stades
square, perimeter	= 31 680 ft.	square, perimeter	= 400 stades
outer circle, radius	= 5040 ft.	outer circle, radius	= $63\frac{1}{2}$ stades
circumference	= 31 680 ft.	circumference	= $399\frac{1}{7}$ stades

That square in conjunction with the outer circular wall provided the Atlantean squared circle.

Like everything in Atlantis its squared circle formula is not quite so neat and accurate as the corresponding figure in Magnesia. A feature in common between the two diagrams is that in both cases the diameter of the central city is equal to the difference between the widths of the square and the circle of equal perimeters. In the geometry of Atlantis, however, there is a slight discrepancy, for the perimeter of the square on the plain in which Atlantis stands is 400 stades, while the circumference of the circular wall is a fraction less, just over 399 stades. A possible means by which the Atlantean geometers could have perfected the squared circle round the city is suggested in the caption to Figure 58, but however they contrived to divide the land on the plain equitably between the 60 000 landed citizens, the result could never have been quite satisfactory. Their cumbersome number system and the ambiguities in their geometry and land measuring would have caused minor grievances which would grow ever more irritating as time went on. In the original formula and foundation plan of Atlantis lay the seeds of its eventual destruction.

In the manner of puzzle-setters Plato used ambiguity and misdirection, and the points on which he withholds information are often as significant as the clues openly provided. A question on which he is blatantly evasive concerns the length of the Atlantean stade. In the dimensions of the circular city the 600-ft. stade is clearly implied, though never actually specified, but the plain could well have been measured by a different yardstick. That brings up the possibility of another version of the squared circle, one in which areas rather than perimeters are made equal. For if the Atlantean plain had been measured in units of the 660-ft. furlong, as used in Magnesia, each of the squares of land formed by the water channels would have been 66 000 rather than 60 000 ft. wide, containing an area of 100 000 acres. As shown above, 100 000 acres is also the amount of land contained within the outer, circular wall of Atlantis, excluding the area of the ringed city. It is possible therefore that the Atlanteans intended to allot 100 000 acres to each group of 100 citizens, in which case every citizen would have 1000 acres. The owner of a farm that size would need a considerable number of retainers. Unlike the Magnesians, who had small plots of land and shared personally all the duties of the state, the landowners of Atlantis were feudal lords, and their service to the community took the form of providing men and equipment from their large estates.

The story of Atlantis, typifying the rise and fall of great civilizations in the past, was in accordance with Plato's own view of history. He believed no doubt that it was based on fact and that it came, as he said, from ancient Egyptian temple records. But the details of its dimensions and so on were clearly of his own devising. To existing traditions of the lost city he attached a mathematical allegory, designed to illustrate the crucial importance of number and true reckoning in all human affairs. The essence of his Atlantis was a numerical pattern which was not supposed ever to have been applied to any actual city. Many of its features are obviously quite impractical, such as the outer wall which overlaps the square of water-channels and encroaches on the neighbouring squares, thus further complicating the problem of land division. That arrangement is simply meant to demonstrate a version of the squared circle. Almost every detail in Plato's account of Atlantis conveys a mathematical clue. They indicate an elaborate geometrical construction, an overblown travesty of the Magnesia pattern though incorporating many of the same sacred numbers and proportions.

The chart of the Atlantis plain could well have been Plato's means of humbling his over-confident pupils by making them work out its various lengths and areas and distribute the land proportionately. Thus he would have ensnared them in a network of numbers and water-channels from which there seemed no escape. At first sight the problem of the plain looks simple, even banal; but investigation proves it extremely complicated, capable of many near-solutions, none of which is ever quite adequate. The effect on Plato's students of working out its proportions would have been to sharpen their wits and introduce them to the type of calculation they would need for appreciating the more sublime figures of sacred geometry such as the plan of Magnesia.

Why Atlantis fell

In Plato's imagery, wealthy Atlantis has the same relationship to simple Magnesia as the great city of Babylon in Revelation has to the New Jerusalem. The two versions of the holy city both derive, as has been shown, from the same traditional diagram and canon of number, and the parallels between the two expansive, mercantile cities which suffered cataclysms, Atlantis and Babylon, are so exact that they also can be assumed to share the same original model. The four cities are really two, and these two relate to each other as opposite reflections of the same image. Though set in contrast as expressions of heaven and hell on earth, they have many features in common, calling to mind those evangelical pictures which show the pathway to paradise alongside the road which leads to perdition.

The foundation plan of Atlantis was in many ways admirable. Poseidon was no mean geometer, and being a god he naturally had access to the divine foundation pattern which ensures the happiness and longevity of communities which adopt it. In its forms and proportions Atlantis closely resembled Magnesia. It was firmly based on the squared circle, the geometer's device for transforming the ideal into the actual, and its numerical framework, featuring the canonical numbers such as 5040, 3168, 144 etc., was essentially that which Plato recommended to the Magnesians. Comparison between the schemes of sacred geometry at the foundations of Magnesia and Atlantis indicates a common origin. Yet Plato claimed that so long as they observed the laws of their founder the Magnesians would

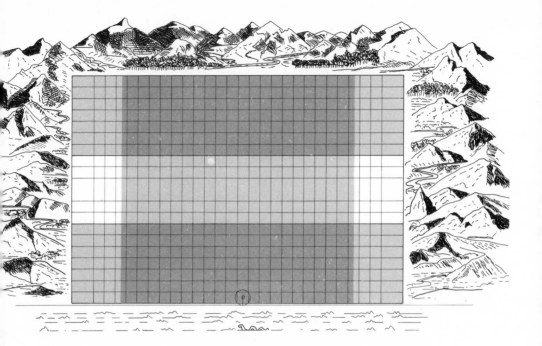

Figure 58. The plain of Atlantis, bordered on three sides by mountains, is enclosed by a rectangular canal which on its southern side runs parallel to the sea. The surface of the plain is divided up by a network of channels. The city lies near the centre of its southern edge. Its circular wall overlaps one of the squares of water channels, producing the squared circle.

For the Atlantean squared circle to be perfect the perimeter of the square made by the water channels must be the same as the measure round the circular wall, $399\frac{1}{2}$ stades. The side of the square must therefore measure, not 60000, but $59871\frac{3}{7}$ ft. This provides a clue to a partial solution of the problem of the plain.

In the diagram above, some of the squares measure the regular 60000 ft. (length a, in white), some have sides equal to that of the square in the Atlantean squared circle, $59871\frac{3}{7}$ ft. (length b, in light grey), and in some both measures are combined (dark grey).

Plato does not say how many channels run across the plain north to south. This allows the striking out of the channels at each end of the row, leaving 27 in all. Thus the total length of the longer side of the plain is made up of:

27 channels each 100 ft. wide =	2700 ft.	
21 × length b.	= 1257300 ft.	
9 × length a.	= 540000 ft.	
Total	= 1800000 ft.	= 3000 stades

Neither the number nor the spacing of the channels running east and west is specified by Plato, so the central channel can be omitted, leaving 18. The shorter sides are made up of:

18 channels each 100 ft. wide =	1800 ft.	
14 × length b.	= 838200 ft.	
6 × length a.	= 360000 ft.	
Total	= 1200000 ft.	= 2000 stades

This cumbersome, awkward arrangement allows the Atlantean squared circle to be perfected, but at the cost of violating several specifications in the foundation plan. Its, decimal, four-square design looks simple, but the fact is that Atlantis does not quite work.

preserve their community from corruption, whereas the Atlanteans, even though they were long faithful to their sacred customs, eventually came to ruin.

With all its advantages of a divine architect and lawgiver, a true cosmological foundation pattern, a fair constitution, a fine climate, excellent soil, natural resources, mineral wealth and all else that should have kept its citizens good and happy, why then did Atlantis fall? The answer is that it was destined to fall from the very start. Poseidon's ambition was the same as that of Magnesia's lawgiver, to make his community as enduring and self-perpetuating as circumstances on earth allow. Unfortunately for the Atlanteans, he began work by making a fundamental error. The mortal element which grew to prevail among them was not merely their inheritance from earth-born Kleito; it arose from Poseidon's original error of choice.

Instead of basing his scheme on duodecimal number and twelve-fold geometry, the founder of Atlantis framed everything in decimals – anticipating a fashionable folly of today. In the symbolism of number, twelve is associated with the gods, the zodiac and the majestic order of the heavens. Being divisible by five out of the first six numbers, it is more complete and provides a more flexible base for numerical compositions than the number ten. That number, which limits the decad and comprehends the tetractys, was revered by the old philosophers for its unique properties, and it is supremely important in the lower branch of mathematics which deals with 'sensibles' and objects of trade and utility. In the higher branch, however, where number assumes its divine function as the first symbol of Creation, the number twelve predominates. It generates all the principal numbers in the canon of music and proportion, and its geometry is far more attractive and simple to develop than the decagonal variety.

It is not easy to construct a decagon or to elaborate a scheme of geometry based on the number ten. Inaccuracies creep in, hard to detect and creating ever widening flaws in the composition, and the decagonal figure is aesthetically not so pleasing as the twelve-sided form. Lacking the four cardinal sides it looks unstable, and it is less inclusive numerically than the New Jerusalem dodecagon, failing to accommodate the geometric forms of numbers three, four, six, eight and twelve which fit naturally into the city of the twelve gates. Poseidon chose to found his settlement on the decimal aspect of the canonical diagram, thinking no doubt that he was being highly practical. The result was to encumber all generations of Atlanteans with a system of geometry and division which soon became complicated and caused friction among the citizens.

In Atlantis nothing quite fits. The geometrical development of its plan, though tolerably accurate to the eye, is not quite precise when examined mathematically. The whole structure was sufficiently well conceived to last many generations, but eventually it had to fall apart.

On the mundane level Atlantis's founder did several things which Plato elsewhere disapproves of. For one thing, he placed his city too near to the sea, exposing its citizens to the corrupting influence of traders, adventurers and foreign tourists. Poseidon being a sea god, it was of course his nature to do such a thing but, according to the notions of Plato, it would have speeded the process of degeneration, causing the Atlanteans to grow dependent on foreign trade, neglect the sensible laws of their founders and develop imperial ambitions. Several features of Atlantis were those forbidden to the Magnesians, its surrounding wall for example, the absence of which in Magnesia was supposed to keep the citizens alert. Yet these were minor and secondary blemishes in comparison with the fundamental error which lay at the very root of Atlantean society and which finally caused the 'mortal element' in its people to prevail over their better natures. The system of number on which all their institutions were based was not the practical, versatile, psycho-therapeutic duodecimal, but the more awkward frame-work of the number ten. It may have been a falling comet that caused the flooding of Atlantis, but however Zeus chose to punish it, the real cause of its downfall was the Atlantean decimal system.

Plato's cosmos and the wanderings of the soul

At the very end of the *Republic* Plato tells an extraordinary story. It is of a type which is strongly represented in folklore and in the literature of psychology and psychical research, about a person on the point of death, or thought to be already dead, who recovers and is able to describe experiences in another world.

Er of Pamphylia was a soldier, left for dead on a battlefield and collected up with the corpses ten days later. The funeral was to be in two days' time, and Er was already lying on the pyre when he came back to life again. The story he told about what happened after his supposed death must have appealed to Plato's imagination, for he adapted it as a parable of the soul's progress through different states of existence and of how the universe is constructed. Er's story was as follows.

When his soul left his body it travelled in company with many others, and they came to a certain weird place where there were two chasms in the earth next to each other, and two other chasms opposite them in the sky. Between the chasms sat judges who, when they had delivered judgment, told the just souls to take the right-hand path and continue on upwards towards the heavens; and they attached marks of their judgment onto them in front. The unjust were ordered to take the left-hand path and proceed downwards; on their backs was the record of everything they had done. When Er came in front of the judges they said that he was to become a messenger to people on earth about the other world, and they told him to hear and observe everything that happened there. He then watched the departure of the souls on whom judgment had been passed, some going into the heavenly opening and some into the opening in the earth. From the other two chasms there came, rising out of the earth, dishevelled and grimy souls and, down from the heavens, souls who were pure and clean. This crowd of arriving souls seemed to have come from a long journey; they were pleased to find a meadow to rest in and they set up a camp there as for a festival. Those who knew each other exchanged greetings, while those who had come from above met those from below and each told of their experiences. Those from below lamented and wept as they remembered all they had suffered and seen during their infernal thousand-year journey, and those from heaven spoke of the incredible beauty and delight of their experience.

There follows an account of the rewards and punishments meted out to each soul in accordance with the nature of its career on earth.

For every wrong a person has committed he must pay the penalty in turn, ten times for each, that is to say, once every hundred years, this being reckoned as the span of a man's life. He pays, therefore ten-fold retribution for each crime . . . and those who have done good and been just and god-fearing are rewarded in the same proportion.

Exceptionally wicked people are confined to the underworld for ever. Whenever they or anyone who has not yet paid the full penalty try to escape upwards, the mouth of the chasm through which they have to pass gives a loud bellow, whereupon they are seized, chastized and thrust down into Tartarus. 'Er said that the fear that the voice would sound for them on the way up was the worst of all the many fears they experienced; and when they were allowed to pass in silence their joy was great.'

Figure 59 illustrates the journey of Er round the universe in company with a party of souls destined for rebirth. He first stands just to the right of the judgment seat, watching the souls of the newly dead arising or descending according to the judgments either by the nearest of the two chasms leading upward or by the opposite, downward chasm. He then steps behind the judges into the meadow and joins the souls from above and below, 'encamped there as for a festival', who are awaiting passage back to earth. They tell of their experiences above or in the nether regions and exchange news of the friends they have met there.

> After seven days spent in the meadow the souls set out again and came on the fourth day to a place from which they could see a shaft of light running straight through earth and heaven, like a pillar, in colour most nearly resembling a rainbow, only brighter and clearer; after a further day's journey they entered it. There in the middle of the light they saw extended from heaven the extremities of its chains; for this light chains the heavens, holding together all the revolving firmament, like the undergirders of men-of-war. And from the extremities they saw extended the spindle of Necessity, by which all the revolving spheres are turned . . .

According to James Adams in his edition of the *Republic* this means that the light passed right through the sphere of the universe, around its axis, and encircled it on the outside. It therefore corresponds to the World-soul which in Plato's *Timaeus* cosmology, examined later, ran through and round about the circle of creation.

The world axis at the centre of the shaft of light is likened in Plato's myth to a spindle. Around it he placed a series of eight rings to represent the orbits of the planets. That aspect of Er's report is considered in the next section. It contains clues for reconstructing Plato's diagram of the universe and allows the hypothesis that the circles of the earth and moon at its centre were given the same dimensions as the corresponding circles in Magnesia and the New Jerusalem diagram. Figure 59 is therefore proposed as an approximation to the diagram which Plato had before him, showing the journey of a soul between two lives. The ring in which Er begins his adventure represents the surface of the earth. Below it is the underworld containing the circle of Tartarus and, in cross-section, the pole of the universe. Above the earth is the sphere of the moon, and the outer circle contains the entire universe. The band of light around the circumference and down the centre of the universe

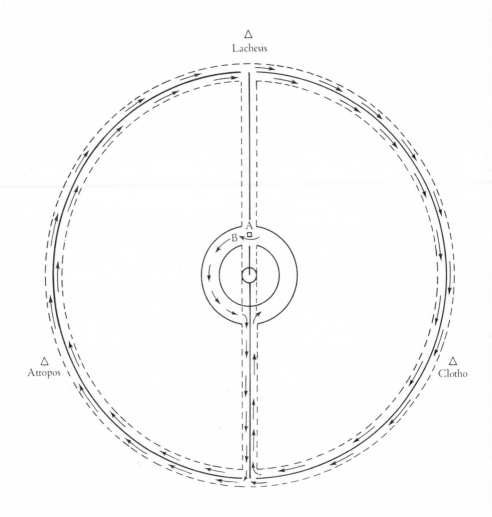

Figure 59. Chart of Er's journey through the other world. Left for dead on the battlefield, he finds himself before the judgment seat (A), then steps behind it onto a meadow (B), meeting the souls who are destined for rebirth. They proceed together as here marked, visiting the three Fates on their way round the outside of the universe before descending to earth. The pole of the universe, which is also perhaps Tartaros, the pit of hell, is at the centre of the diagram, and around it are the nether world, the surface of the earth and the lunar sphere. The outer circle contains the whole universe.

is shown by broken lines. It is the element through which the souls travel.

After their encampment on the meadow, the souls to be reborn, in company with Er, are shown in the diagram as taking the higher of the two passages above the surface of the earth, travelling through that region which in the New Jerusalem plan is the upper limb of the moon. Having completed a semicircle, they found themselves within the shaft of light, gazing down its axis. This took place on the fifth day or about $4\frac{1}{2}$ days after the start; so, counting the seven days spent on the meadow, Er had been with the souls for $11\frac{1}{2}$ days, leaving only about half a day before his body was due to be cremated. Subsequent events must therefore occur almost instantaneously if he is to resume his career on earth, and this is possible, because the rest of the journey takes place in the spiritual realm of the light beyond the limits of time.

According to the interpretation given in figure 59, Er and the souls are transported in a flash down the shaft of light to the bottom of the diagram, and then clockwise round half the circumference of the universe to its highest point, where they encounter Lachesis. She is one of the three Fates, daughters of Necessity, who sit enthroned at equal intervals round the outer circle of the universe, turning the rings of the spindle.

In the presence of Lachesis the souls drew lots to decide in which order they were to choose future lives from a large selection laid out on the ground in front of them. Plato lingers over this scene, providing some amusing anecdotes to illustrate the workings of metempsychosis. This is the moment, he says, which justifies a lifetime spent in the study of philosophy, for one's whole future depends on how wisely one chooses. The general rule is that people who have enjoyed orderly lives on earth and the delights of heavenly existence are less cautious than those who have recently emerged from torment. They are liable to choose the life of a great dictator or a glamorous career which ends disastrously and leads them into the nether regions. In contrast, the hero Odysseus, tired of adventures and suffering, picked for his next round the life of a quiet, ordinary citizen. It was a life which those before him had spurned, but Odysseus said that if he had had first pick that would still have been his choice. Others continued the habits of their previous incarnations. Thus Atalanta elected to become a great athlete, and a wellknown buffoon took the life of an ape. Orpheus, who hated women because he had been killed by them, wanted to be a swan, Thamyris a nightingale and Ajax a lion. 'And there were many other changes, from beast to man and from one kind of beast to another, the unjust becoming wild animals and the just tame in all sorts of interchanges.'

When each soul had selected the kind of life he wanted he was allotted a guardian angel to help him fulfil his choice, and the party continued on round a third part of the circle to the throne of the next Fate, Clotho, where their decisions were formally confirmed. Er alone, not having drawn a new life, was exempt from these proceedings. Another third of the circle brought them all before the third Fate, Atropos, who spun the threads of their destinies so as to make them irreversible.

And then, without turning back, each soul came before the throne of Necessity, and passing before it waited till all the others had done the same, when they proceeded together to the plain of Lethe through a terrible and stifling heat; for the land was without trees or any vegetation. In the evening they encamped by the River of Oblivion, whose water no pitcher can hold. And all were compelled to drink a certain measure of its water; and those who had no wisdom to save them drank more than the measure. And as each man drank he forgot everything. They then went to sleep and when midnight came there was an earthquake and thunder, and like shooting stars they were all swept suddenly up and away to be born. Er himself was forbidden to drink, and could not tell by what manner of means he returned to his body; but suddenly he opened his eyes and it was dawn and he was lying on the pyre.

The diagram shows the positions of the three Fates and the course followed by Er and his companions in visiting them in turn. It appears, since they are not allowed to turn back, that they have to go twice round the outside of the universe before arriving at the throne of Necessity, the lowest point of the diagram. From there they plunge up the right-hand channel of the shaft of light, passing through the torrid, treeless regions of the planets to the banks of the River Oblivion, which is probably located where the shaft of light enters the first of the inner circles. Having drunk of its waters, they are carried down to the surface of the earth to be reborn. The entire journey, says Plato, presumably meaning the journey from one birth to the next, lasts a thousand years. In the case of Er, who did not have to experience that part of it lying between judgment and the gathering of souls on the meadow prior to rebirth, it took twelve days.

There is still a large gap in the diagram, between the sublunary world at its centre and the highest heavens. This has now to be filled with the planetary orbits as described in the story of Er.

On their journey from the meadow to the thrones of the Fates the souls returning to earth entered the light which streams through the universe and looked along its axis which resembled the shaft or spindle of a spinning-wheel. Around it there revolved a series of *sphonduloi* or 'whorls', turning upon the spindle like a nest of bowls.

There were eight whorls in all, fitting one inside the other and showing their rims on the surface like so many circles, so that they formed a single whorl with a continuous surface around the shaft which is driven right through the middle of the eighth.

The surfaces of the whorls thus form a continuous plane as eight concentric rings round a shaft. Details are then given of their relative widths, colour tones, speeds and types of motion.

The first and outermost whorl had the broadest rim; next broadest was the sixth, then the fourth, then the eighth, then the seventh, then the fifth, then the third, and narrowest of all was the second.

The rim of the largest and outermost was of many colours, the seventh was the brightest, the eighth received its light from the seventh, and also its colour, the second and fifth were like each other and yellower than the rest, the third was the whitest, the fourth reddish and the sixth second in whiteness.

The whole spindle revolved with a single motion, but within the movement of the whole the seven inner circles revolved slowly in an opposite direction to that of the whole; and the eighth moved fastest, the seventh, sixth and fifth moved at the same speed and were the next fastest, the fourth whorl was the third in speed, moving as it appeared to them with a counter-revolution, fourth in speed was the third and in fifth place was the second. And the whole spindle turns in the lap of Necessity. And on top of each circle stands a siren, which is carried round with it and utters a note of constant pitch, and the eight notes together make up a single harmony.

The music of the sirens is echoed by the three Fates. Lachesis sings of things past, Clotho of things present and Atropos of things to come. The Fates also help turn the whorls of the spindle. Clotho sometimes turns the outermost rim, Atropos turns the inner rim with her left hand, while Lachesis turns the inner and outer rims with her left and right hand alternately. The hand used by Clotho is not mentioned, but for the arrangement to be symmetrical it should be her right, and for that reason the outer rim in the diagram of Er's journey has been given a clockwise motion.

The given attributes of the eight whorls and their supposed astronomical symbolism are set out below, number 1 being the outermost.

number of whorl	order of width	order of speed	colour tone	astronomy
1	1	6	many-coloured	fixed stars
2	8	5	yellowish	Saturn
3	7	4	whitest	Jupiter
4	3	3	reddish	Mars
5	6 ⎫		yellowish	Mercury
6	2 ⎬ 2		second whitest	Venus
7	5 ⎭		brightest	Sun
8	4	1	illuminated by 7	Moon

The problem is to resolve these data into a coherent scheme of music and astronomy. Each of the whorls contained a planet, carried round with the circular motion of the whorl, and on each planet was perched a siren, singing her particular note. If the planets fitted the whorls containing them so that the diameter of the planet was the same as the width of its whorl, each planet would be tangent to its neighbour whenever they passed each other in orbit, which is how the earth and the moon are placed in the New Jerusalem diagram. The diameters of the planets can be seen as corresponding to the strings of a lyre, in which case the note sung by a siren would be in accordance with the length of diameter of the planet she rode upon. Many of the scholars who have commented on the story of the whorls have assumed that the harmony made by the singing of the eight sirens consisted of the eight notes of the octave. This is neither stated nor implied in Plato's text, and it is unlikely that he would have taken the range of a single octave to encompass the entire music of the spheres. The canon of 'lawful' music on earth was meant to echo the divine music of the heavens, and since the human voice can range over at least two octaves, that range must be attributed also to its heavenly model. The song of the sirens could scarcely have been more limited in its range than the voice of the human singer who was supposed to imitate it.

Adrastus, quoted by Theon of Smyrna, reported that the fourth century BC Greek philosopher, Aristoxenus, 'limited the extension of the diagram which represents the different modes to the double octave and the fourth' – according to Hiller's Greek text of 1878. In the recent translation by R. and D. Lawlor from J. Dupuis's French edition of 1892 this is rendered 'two octaves and a fifth'. The difference between a fourth, 3:4, and a fifth, 2:3, is a tone of 8:9 which the ancients took as the smallest interval clearly distinguishable between two sung notes. Two octaves and a tone was the range of the Greek pentedecacord, the fifteen-stringed lyre. There was evidently disagreement on how to define the limits of the human voice, though by common consent it encompasses rather more than two octaves. Taking the greater value of two octaves and a fifth, the canonical range of the voice is expressed by four numbers, 3:6::9:18, consisting of two octaves, 3:6 and 9:18, plus a fifth, 6:9.

'Greek musical theory', says Ernest McClain (*The Pythagorean Plato*), 'is founded on the so-called "musical proportion" 6:8::9:12 which Pythagoras reputedly brought home from Babylon'. The octave 6:12 is made up of a fourth (6:8), a tone (8:9) and a fourth (9:12). With this division of the octave, the two octaves and a fifth making the canonical range of the human voice are represented in simplest terms by the eight numbers, 6, 8, 9, 12, 18, 24, 27, 36, giving the musical sequence:

$$6 \times 4/3 = 8$$
(fourth)
$$8 \times 9/8 = 9$$
(tone)
$$9 \times 4/3 = 12$$
(fourth)
$$12 \times 3/2 = 18$$
(fifth)
$$18 \times 4/3 = 24$$
(fourth)
$$24 \times 9/8 = 27$$
(tone)
$$27 \times 4/3 = 36$$
(fourth)
$$36$$

These numbers give the proportions of the eight rings or planetary whorls surrounding the shaft of the spindle. The shaft itself, if it is to be in harmony with the whorls it turns, must be less than the first number in the series by the ratio of a musical fifth. The first number being 6, the radius of the shaft is therefore $6 \times 2/3 = 4$. Adding 4 to the sum of the other eight numbers gives the full radius from the centre of the spindle to the extremity of the outer whorl, which proves to be the appropriate number, 144.

Figure 60 shows the plan of the whorls, arranged in the order given by Plato and proportioned in accordance with the above numbers. From the central shaft, radius 4, the widths of the successive rings are 18, 12, 27, 9, 24, 8, 6, 36. The proportions of the two inner circles round the hub are the same as those of the earth and moon circles in the New Jerusalem, and the same dimensions can be obtained by multiplying all the terms in the series by 180. The widths of the rings round the shaft (radius 720) are thus made, 3240, 2160, 4860, 1620, 4320, 1440, 1080, 6480. The total width of the eight whorls together is 25 200, and the radius of the circle containing them all, together with the shaft, is 25 920. Omitting the central shaft, the area of the circle formed by the whorls combines two important symbolic numbers, being equal to $31 680 \times 666$.

The chart of Er's journey and vision of the universe is now complete. At its centre is the circular cross-section of the universal pole or the spindle which turns the whorls, radius 720 ft.

The next outer ring has a width of 3240, making its radius from the centre $720 + 3240 = 3960$. This circle represents the earth, radius 3960 miles, as well as the circle of cultivated land, radius 3960 ft., in Magnesia.

The next larger circle has a radius of $3960 + 2160 = 6120$, which is the same as that of the outer circle in the New Jerusalem. The width of this ring in section being 2160, it contains the circle of the moon, diameter 2160 miles.

Most authorities locate the moon and the sun in these inner rings, numbers 7 and 8, because ring 8 is said to borrow its light from ring 7 as the sun illuminates the moon. But this may be one of Plato's red herrings. His description of ring 7 as 'brightest' and as giving light to ring 8 is obviously applicable to the sun in relation to the moon, but equally the earth is illuminated by the moon, and at night the moon is the brightest object in the sky. There is reason, therefore, to be doubtful of the generally accepted order of the planetary whorls, as listed on page 144. Yet, however one interprets it, Plato's scheme of the heavens seems greatly at odds with the actual facts of

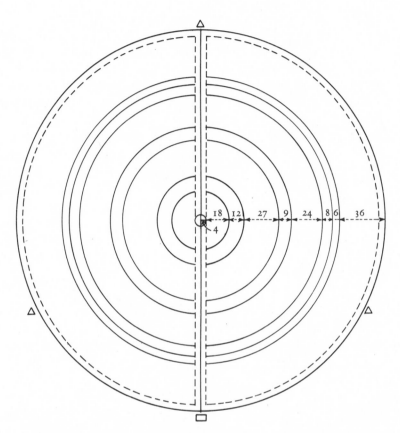

Figure 60. The plan of the heavenly whorls as revealed to Er. They are placed in the order specified by Plato, and their widths are given in the simplest terms. Multiplied by 180, the dimensions exhibit the canonical numbers of Platonic arithmetic, and the rings of the New Jerusalem are discovered at the centre of the diagram.

astronomy. His apparent description of the solar system refers symbolically to the whole universe and was clearly not intended as a picture of any particular phenomena within it. It represents the ideal, archetypal universe which Plato recommended as the proper study for astronomers, advising them to 'look down, not up'. In Plato's system the essence of nature was number, so his ideal cosmology was a comprehensive pattern of numerical harmony, the central core of which he expressed geometrically as the city plan of Magnesia. Being also the traditional image of the sublunary world at the heart of the universe, that plan becomes the centre-piece of the symbolic

Figure 61. The plan of the heavenly whorls and of the soul's wanderings between incarnations on earth can be envisaged as a labyrinth, ancient symbol of the path through life and death. The rings of Plato's diagram are here adapted to the traditional labyrinth form, the souls of the dead proceeding upwards or downwards through the spirals.

world order developed in Figure 60. Featured in it are the same proportions, musical harmonies and symbolic numbers as appear in the later study of the harmonious World-soul in the *Timaeus*.

The chart of the soul's progress from death to rebirth can also be seen in the traditional form of a giant labyrinth, embracing both the subterranean world and the rings of the heavens (figure 61). Plato several times hints at such a figure. In *Phaedo* (111–114) he describes the regions below earth as a labyrinth of rivers and channels with Tartarus, where the truly wicked are confined for ever, at its centre. Another labyrinth or series of rings is mentioned in *Phaedrus* (247–248) as constituting the heavenly realm. The gods in their chariots can easily ascend to the highest ring, and souls who achieve immortality can follow them there. 'For those that are called immortal, when they reach the top, pass outside and take their place on the outer surface of the heaven, and when they have taken their stand, the revolution carries them round and they behold the things outside of the heaven.'

To that region of pure intelligence, farthest removed from mundane illusions, the philosophic soul can aspire by completing three exemplary lives on earth; but few are able to rise so high. The great majority of souls who are permitted after judgment to take the path leading upward find themselves weighed down by the earthly part of their nature and can only soar to one of the lower rings of heaven, from which the world of divine truth is but dimly glimpsed. After their period above they fall back into rebirth.

In the myth of Er Plato tells a good story while imparting under the veil of allegory the traditional code of number behind the order of the universe and the canon of music. At the same time he instructs on the benefits of philosophy, the advantages of leading a good, moderate life and the fate of the soul after death. Thus he concludes the *Republic* with a parable summarizing the important features of his doctrine, both the esoteric and the openly proclaimed. Of most immediate interest to everyone is his account of the soul's passage through the labyrinth of the after-world and how best to negotiate it. The crucial moment is when it comes to choosing the pattern of one's next life. Once that choice has been made it is unalterable, one's career proceeds accordingly and rewards or punishments duly follow. The best way of spending this life, says Plato, is in cultivating sound philosophy and the habits of justice, so that when the time comes to pick a new career one is trained to make a wise choice. We are not told which form of career is the most favourable, but that the choice should be governed by love of justice which leads the soul upwards. Specifically to be avoided are:

> the temptations of wealth or other evils, and descending to the life of a tyrant or some other type of malefactor, committing intolerable evils and suffering worse oneself. Rather we should learn how to choose the moderate path and avoid as far as possible, in this life and the next, the extremes on either side. For this is the surest way to human happiness.

Plato clearly believed in what he preached, the immortality of the soul and the doctrine of metempsychosis. His studies in the Orphic mysteries had probably persuaded him that these things were true, and he was evidently impressed, as are many people today, by stories such as Er's. He also recognized the Pythagorean belief in techniques for remembering past lives; hence his advice to drink sparingly of the waters of the River Oblivion. Although many of the details in the journey of Er were obviously not meant literally, but refer to the geometric diagram which illustrates it, the general pattern of events was in accordance with Plato's own convictions. His recipe for a successful cosmic career is said also to be that which produces the greatest amount of happiness in this present life. It is:

> to believe the soul to be immortal, capable of enduring all evil and all good, and always to keep our feet on the upward way, pursuing justice and wisdom. Thus shall we be at peace with God and with ourselves, both in our life here and when, like victorious athletes collecting their prizes, we receive our rewards; and in both this life and the thousand-year journey which I have described all will go well with us.

The numerical creation myth of 'Timaeus'

In the *Timaeus* Plato describes the cause and processes of universal creation. The story is told simultaneously in two forms, mathematical and mythical, the first appealing to reason and intellect, the second supplying imagery and humour. Written towards the end of his life, the *Timaeus* was Plato's most powerful achievement. It provided a cosmology which sustained western civilization for some two thousand years and has been a constant source of renaissance. Many of its illustrative, inessential details may be disputed (e.g. Plato's idea that the souls in women are those formerly in cowardly and immoral men), but the answers it gives to those questions which arise with consciousness, such as the purpose and origin of the world, its true nature, its form and limits and whether or not it is immortal, are coherent and reasonable. Its survival and regularly renewed influence confirm Plato's own view, that it is the 'most likely' or best possible image of reality, producing the best results. The traditional cosmology transcends and does not conflict with the proofs of modern physics, and when the necessity for a new philosophical standard for the humane development of science becomes apparent, the *Timaeus* offers the guiding principles for its constitution.

The best reason one can find for the creation of the universe, says Plato, is that the eternal God who experiences constant perfection wished out of kindness to share his state as widely as possible. He embarked therefore on his unique work, a living creature modelled on the pattern of eternity and closely resembling it. The creature's shape was a perfect sphere, smooth on the outside and with no external organs or appendages. These were unnecessary, since the universal globe contained everything within itself and there was nothing and nowhere beyond it. It revolved on its axis but was otherwise motionless in the void. Since there was nothing to erode or corrupt it from outside, it was almost immortal; but it was not entirely so because, being only a model of eternity, it could not fully share in the quality of the original.

The microcosm of the universe, the human head, was also made spherical. As a secondary form, subject to outside influences and different types of motion, it lacked perfect symmetry and, since it could not well be left to roll about the earth, a corporeal carriage and limbs were added onto it. Thus the body was enabled to 'pass through every place, bearing on high the head, our most divine and sacred habitation'. Organs of perception were placed on one side of the sphere so that it should move mostly forwards, which is more dignified than backward motion.

To explain the appearance of the three principal kinds of animals which inhabit, in descending order of merit, air, earth and water, Plato adopted an evolution myth which was the exact opposite of Darwin's – epitomizing the difference between ancient and modern philosophy. His one point of agreement with the moderns was that 'animals have always been changing one into the other', but he reversed the now-accepted process by deriving lower creatures from degenerate men. Birds were those who had once been astronomers of the type Plato scorned, who believe that real knowledge can be acquired by studying the physical form of the heavens. They were given wings and feathers and made to expiate their folly in those upper regions which had once allured them. Land animals were men who had ignored philosophy and never raised their sight above the mundane level; because of which they were tied to the earth by four or more legs and, in the worst cases, crept along its surface as reptiles. The grossest of former men were not even allowed to breathe fresh air but were confined below the waters in the forms of fish or molluscs.

Plato was obviously no Darwinian, but neither did he suscribe to any crude fundamentalist form of creationism. The ineffable God of the universe did not directly fashion the various types of creatures within it. That was the task of a hierarchy of lesser gods or creative powers, divinely appointed to create and maintain order. The materials they used were coarser compounds than those of the primary creation, and the results of their handiwork were less than perfect. Under their guidance, however, the internal functions of the universal organism are perfectly balanced, and the whole system is designed to perpetuate itself as long as God so wills.

With many other pleasant fables Plato described the living contents of the universe, particularly its human element, both physically and psychologically. Many of his notions have an odd ring and seem barely credible today, but they all contribute in some way to the world-view which Plato considered to be the best foundation for humane societies. Thomas Taylor, the most philosophically instructed of recent Platonists, claimed that 'almost every word (in Plato) has a peculiar signification and contains some latent philosophic truth'. He further asserted that whoever studies the theory of the universe set out in the *Timaeus* will be convinced 'how infinitely superior the long lost philosophy of Pythagoras and Plato is to the *experimental farrago* of the moderns'.

While spinning his amusing tales about the nature of the universe, often in the same words, Plato gave a scientific account of creation with mathematical formulas for the benefit of his initiated readers. His Creator proceeded in the same way as God in the Apocrypha (Wisdom, 11,20) of whom it is said 'thou hast ordered all things in number and measure and weight'. As the first stage in his plan to make all things as like himself as possible, he reduced primeval chaos to order by setting numerical proportions between its elements. These were symbolically named earth, water, air and fire. The Creator fabricated the universe from the most and the least corporeal of the elements, earth and fire, linking them by placing water and air as mean terms in a geometric progression which bound them all together.

In bringing order to the universe the Creator gave geometric forms to the atoms of which the four elements are separately composed. The simplest of the five regular figures of solid geometry, the tetrahedron, is the shape of the fiery atom, fire being the most volatile of the elements and the tetrahedron the least stable of the solids. The most stable, the cube, forms the atoms of earth, the octahedron is the form of air and the twenty-sided icosahedron of water. Compounded together in various proportions and of different sizes, the atoms of the four elements make up the entire physical universe, interacting with human senses to produce appearances and feelings. The fire atom, for example, which has the sharpest angles and edges, causes the sensation of heat by cutting and stinging.

The Platonic solids, so called because of their appearance in the *Timaeus*, are there defined as 'solid figures which divide the surface of a circumscribed sphere into equal and similar parts'. There are only five of them, those which Plato related to the four elements together with the dodecahedron which, he said, 'was used by God for arranging the constellations on the whole heaven'. Elsewhere (*Phaedo*, 110) Plato compares the earth to a dodecahedron, saying that when seen from above it resembles 'one of those balls which are covered with twelve pieces of leather'.

In the course of their interactions and collisions atoms are constantly being broken up, whereupon their constituent parts come together again, sometimes forming different bodies and being transformed from one kind of atom into another. These transformations occur between the atoms of fire, air and water because the solids which represent them, the tetrahedron, the octahedron and the icosahedron, are each made up superficially of the same kind of triangle, the equilateral. When the atoms are broken up, the triangles into which they fall reassemble to make new atoms of either fire, air or water.

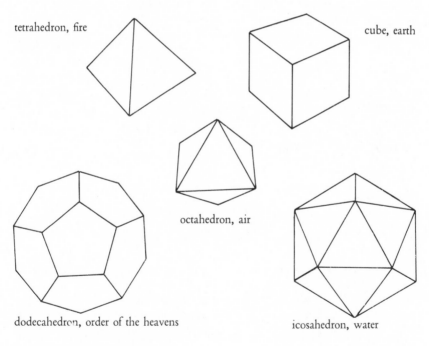

tetrahedron, fire

cube, earth

octahedron, air

dodecahedron, order of the heavens

icosahedron, water

Figure 62. The five Platonic solids.

The cube, which is the solid representing earth, can not however be divided into equilateral triangles; the paired triangles which make up each of its sides are isosceles. When broken up, the earth atoms do not therefore transform into other types of atom but reform themselves as cubes.

A model of the physical universe, as Plato described it, would be a perfect, all-inclusive sphere, containing every type of proportion as represented by the five regular figures of solid geometry. Such a model was constructed in the sixteenth century by Johannes Kepler, demonstrating celestial mechanics by a combination of solids containing the spheres of planetary orbits, all within an outer sphere beyond the fixed stars. The universe, however, is no mere mechanism but a living organism by virtue of its soul. According to Plato the World-soul was the Creator's primary work, and in *Timaeus*, 35–36, he gave the formula of its composition.

The harmonic composition of the World-soul

In describing the process by which the Creator fabricated that quality called Soul which gives life to the world, Plato was at his most explicit on the nature of Pythagorean arithmetic and the traditional canon of number. The World-soul was compounded from three ingredients, the two opposites called the Same and the Other together with Essence. These were themselves compounds, being made up of their two aspects, the eternal and the transient. They were blended together in a bowl, and the Creator then proceeded to cut off seven portions of the mixture, each of a different size, as follows.

> First he took one portion from the whole;
> then he took a double portion;
> then a third portion half as much again as the second, or three times as great as the first portion;
> the fourth portion he took was twice as great as the second;
> the fifth portion was three times as great as the third;
> the sixth was eight times as great as the first;
> the seventh was twenty-seven times as great as the first.

It is not said what proportion of the whole mixture was represented by the parts thus taken from it, but the proportions between those parts were as 1, 2, 3, 4, 9, 8, 27, consisting of the double intervals, 1, 2, 4, 8, and the triple intervals, 1, 3, 9, 27.

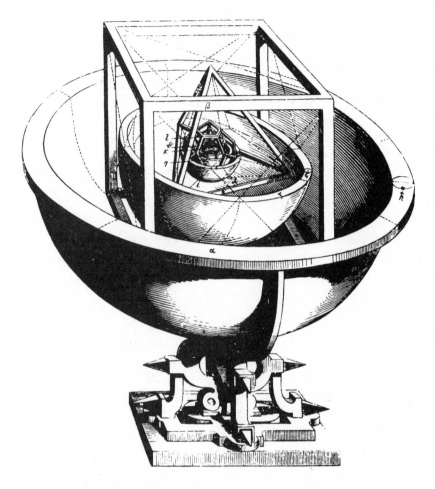

Figure 63. Kepler's model of the five solids fitting together within the sphere of the universe.

The Creator went on to fill the gaps in the series of double and triple intervals by cutting off further portions of the Soul material so as to provide two means between each successive pair of terms. The two means were the arithmetical and the harmonic, as defined on page 90.

The insertion of these links formed new intervals in the previous intervals, those of 3:2, 4:3 and 9:8. The Creator filled up the 4:3 intervals with those of 9:8. This still left over in each case a fraction represented by the terms of the numerical ratio 256:243.

And thus the mixture from which he had been cutting off all these portions was now all used up.

This numerical operation is more fully described below. As the result of it the Creator was left with a strip of Soul material, made up of segments in the proportion of a musical scale. This he divided into two by cutting it lengthwise down the centre. He then laid the two strips one over the other to form a cross, and bent both strips into circles, fastening them together in two places so that one circle contained the other. The whole figure was made to revolve on its axis, but the two bands turned in opposite directions, the outer having the 'motion of the Same' and spinning 'to the right along the side', while the inner was given the 'motion of the Other' and turned 'to the left along the diagonal'.

The circle of the Same remained perfect and entire, but the inner circle of the Other was 'split in six places into seven unequal circles according to each of the double and triple intervals, three double and three triple' – that is, in the proportions 1, 2, 4, 8, and 1, 3, 9, 27. Three of these were made to revolve at the same speed and the four others at different speeds, 'the ratios of which one to another are those of natural integers'.

Having made the World-soul and 'woven it throughout the heavens every way from the centre to the extremity, enveloping it as a circle on the outside', the Creator shaped the physical world and fastened it together with the Soul structure, 'joining them centre to centre'. The seven circles into which he had split the circle of the Other provided the orbits of the heavenly bodies, of which five are named: the earth at the centre, the moon, the sun, Venus and Mercury. In their movements was the origin of Time, which is subservient to number, and thus the numerical essence of the universe was made apparent. Above all there is one particular number which, says Plato, is that of the Complete Year in which all the planets fulfil their orbits and return to the same position relative to the circle of the Same.

The difference between the created universe and the perfect pattern on which it was designed is the difference between an original and a copy. One of the basic tenets of Plato's philosophy was that a copy is bound to be inferior to its model. Thus God's creation was necessarily less perfect that the state of eternity which it imitated. It is subject to Time, and all its parts are transient, 'continually being about to exist, existing and having existed'. Plato summed up the universe in one phrase, as a 'moving image of eternity'.

For over two thousand years Plato scholars have laboured inconclusively to establish the exact numerical composition of his World-soul; for therein is a great prize. Encoded in the *Timaeus's* brief history of creation is the essential knowledge which Plato distilled through his study of traditional science. Among its contents are the canon of musical harmony which determined the forms of government in the ancient world, the scientific cosmology which gave it stability and, above all, the basic formula behind that mathematical standard from which ancient philosophy and every branch of art and science were derived and developed. That is the prize which has tempted the learned in every age to try their hand at decoding the *Timaeus*. If ever it were to be gained, the world would not of course be changed overnight; but the recovery of the ancient cosmic standard would remove from idealism the stigma of impracticality; and if ever there should arise a ruler in the Platonic mould, aspiring to become that philosopher-king whom Plato believed to be the only possible instrument of fair, humane government, such a ruler would have the benefit of a constant mathematical standard as guide and reference in all things.

Yet the returns so far obtained from investigation of the Platonic number code are disappointingly slender when compared with the amount of effort which has been put into it. The history of the problem, from its discussion in Plato's Academy after his death to the modern essays of A.E. Taylor and the musicologist Ernest McClain, shows little evidence of progress. Some of the most important contributions to it are recapitulated below, and some conjectures are added about the content of that musical scale, set out as a progression of numbers, by which the Creator fashioned the World-soul.

The mathematics of creation begins with two numerical progressions. That which makes up the physical world consists of four numbers representing the four elements, arranged in a geometrical progression so that the least substantial element, fire, is linked to the densest, earth, by means of air and water. The formula is, fire : air :: air : water :: water : earth. For the World-soul another progression was used, probably the harmonic type, the terms in which were the qualities called Same, Other and Essence. The problem here is to identify the numbers symbolized by the four elements and the three qualities, and thence, by adding together the numbers of the four elements, to discover the aggregate number of the physical world, and by adding those of the qualities to find the number of the World-soul.

Plato gives no apparent clues as to the numbers comprising the physical world, but its aggregate number is likely to be the same as that of the World-soul since the two were designed to fit together. The aggregate number of Soul is indicated in the description of how the Creator portioned out its material so as to form the strip which he later bent into circles. Since he used up all the Soul material by removing variously proportioned parts of it, the sum of the numbers represented by those parts must be equal to the total number of Soul. If those component numbers could be identified the recovery of the traditional formulas of creation would be virtually accomplished.

Of the many scholars who have offered solutions to the problem of the World-soul scarcely any two have arrived at exactly the same figures. The following analysis proceeds from firm ground, beginning with the original divisions of the Soul material which Plato specifies, into the areas where disagreements arise.

The traditional mode of exhibiting the seven basic numbers of the World-soul, as practised in Plato's Academy, is in the form of the Greek *lambda*, the letter L, thus:

$$\begin{array}{ccc} & 1 & \\ 2 & & 3 \\ 4 & & 9 \\ 8 & & 27 \end{array}$$

The left leg of the *lambda* represents the double intervals and the right the triple intervals. Each of these intervals must now be filled with an arithmetical mean and a harmonic mean. The Pythagorean convention is against the use of fractions on the grounds that unity is indivisible, so in order to obtain whole numbers for the means to be inserted it is necessary to multiply the basic numbers by 6. The three intervals in both series being each filled by two means (italicized below), the result is two series of ten numbers:

6, *8*, *9*, 12, *16*, *18*, 24, *32*, *36*, 48 for the double intervals;
6, *9*, 12, 18, *27*, *36*, *54*, *81*, *108*, 162 for the triple intervals.

Between the successive terms of these series there are, as Plato says, intervals of 3:4, 4:3 and 9:8. His next instruction is to divide the 4:3 intervals, which are those of a musical fourth, into intervals of 9:8, which is the value of a musical tone. Two tones do not quite amount to a fourth; left over is the interval 256:243 which is known as the leimma. This is shown by the equation, $9/8 \times 9/8 \times 256/243 = 4/3$. For the 4:3 intervals to be divided in this way so that every term in the series remains an integer, the series must be further multiplied, and at a later stage they have to be increased again, the necessary multiple being 64. The original terms, 1, 2, 3, 4, 8, 9, 27, have thus to be multiplied by 6×64 or 384, so the first term in the series becomes 384 and the last is $27 \times 384 = 10\,368$. This makes the sum of the seven original numbers, each having been multiplied by 384, equal to 20736 or 12^4. The list below is of the original terms in the series, together with the harmonic and arithmetical means between them, after all have been multiplied by 384.

double series			triple series		
term	term × 384	interval	term	term × 384	interval
1	384		1	384	
		— 4:3			— 3:2
h.m.	512		h.m.	576	
		— 9:8			— 4:3
a.m.	576		a.m.	768	
		— 4:3			— 3:2
2	768		3	1152	
		— 4:3			— 3:2
h.m.	1024		h.m.	1728	
		— 9:8			— 4:3
a.m.	1152		a.m.	2304	
		— 4:3			— 3:2
4	1536		9	3456	
		— 4:3			— 3:2
h.m.	2048		h.m.	5184	
		— 9:8			— 4:3
a.m.	2304		a.m.	6912	
		— 4:3			— 3:2
8	3072		27	10368	

The 4:3 intervals are now filled up by intervals of 9:8, 9:8, 256:243, producing the following series:

double series		triple series	
number	interval	number	interval
384		384	
	9:8		3:2
432		576	
	9:8		9:8
486		648	
	256:243		9:8
512		729	
	9:8		256:243
576		768	
	9:8		3:2
648		1152	
	9:8		3:2
729		1728	
	256:243		9:8
768		1944	
	9:8		9:8
864		2187	
	9:8		256:243
972		2304	
	256:243		3:2
1024		3456	
	9:8		3:2
1152		5184	
	9:8		9:8
1296		5832	
	9:8		9:8
1458		6561	
	256:243		256:243
1536		6912	
	9:8		3:2
1728		10368	
	9:8		
1944			
	256:243		
2048			
	9:8		
2304			
	9:8		
2592			
	9:8		
2916			
	256:243		
3072			

The first series now represents a musical scale in the traditional Dorian mode which Plato recommended in the *Republic* as having the most beneficial effects on societies and the individual soul. It extends over three octaves and forms a regular progression, tone, tone, leimma, tone, tone, tone, leimma, thrice repeated. The second series, however, has gaps in the form of intervals of 3:2, the musical fifth. Plato does not specify how these intervals are to be filled, nor does he say how the two series are to be brought together so as to form the single strip which the Creator made from the various portions of Soul material. In the first series the 3:2 interval between the first term, 384, and the fifth term, 576, is filled by three tones and a leimma in the order 9/8, 9/8, 256/243, 9/8, making the sequence 384, 432, 486, 512, 576, and it seems likely that each of the 3:2 intervals in the second series was similarly filled. The 'fillings' would then be:

between 384 and 576:	432,	486,	512
between 768 and 1152:	864,	972,	1024
between 1152 and 1728:	1296,	1458,	1536
between 2304 and 3456:	2592,	2916,	3072
between 3456 and 5184:	3888,	4374,	4608
between 6912 and 10368:	7776,	8748,	9216

When these eighteen new terms are added, the second series is found to contain all the terms which appear in the first series except for the number 2048. If this is included, the full range of numbers constituting the World-soul appears to be:

	384	768	1536	3072
$9:8$ —				
	432	864	1728	3456	6912
$9:8$ —				
	486	972	1944	3888	7776
$256:243$ —				
	512	1024	2048		
$3^7:2^{11}$ —				
			2187	4374	8748
$256:243$ —				
	576	1152	2304	4608	9216
$9:8$ —				
	648	1296	2592	5148	10386
$9:8$ —				
	729	1458	2916	5832	
$9:8$ —				
			6561		

The numbers of the World-soul are here arranged to be read down the columns, each number being double the corresponding number in the preceding column. There are 35 numbers in all and their total sum is 108 551.

This reckoning agrees with that of J. Dupuis, set out in a note to his 1892 French translation of Theon of Smyrna. Yet, though carried out in strict accordance with Plato's instructions, it is almost certainly deficient. Neither the number of terms, 35, nor their sum, 108 551, is of any particular mathematical importance, and there is no reason why Plato should have called attention to either number. Most of the numbers on the above list have been accepted by all authorities, but a review of the historical solutions to the problem shows how wide has been the range of disagreement, even from the earliest times, as to the number of terms comprising the World-soul and their aggregate number.

The *Timaeus Lochris*, a Pythagorean work thought to be of the first century AD, has a list of 36 terms, adding 6144 or 384×16 to the numbers given above, which makes their total $108 551 + 6144 = 114 695$. In his fifth-century *Commentaries on the Timaeus* Proclus summed up previous essays on the numbers of the World-soul and offered a solution of his own, rejecting from the *Timaeus Lochris* list the numbers 2187 and 6561 and making the number of terms 34 with an aggregate of 105 947. The great nineteenth-century Platonic editor, A.E. Taylor, included 6144 and added two other terms, 4096 and 8192, while omitting the terms 2187, 4374, 6561 and 8748. Thus he arrived at a total of 34 terms amounting to 105 113. The most recent contribution, by Ernest McClain in his book, *The Pythagorean Plato*, revives a suggestion by the ancient writer, Severus, recorded by Proclus, that the commonly accepted numbers of the Soul should be doubled, starting with 768 instead of 384 and ending with 20 736 rather than 10 368. The number of terms he proposes is 41. McClain's reasons for this arise from his expertise in ancient musical theory, which makes his book essential reading for students of this problem. It is not doubted that Plato intended the World-soul to form a musical scale extending over four octaves and a major sixth. Yet that was not his only intention. In accordance with the Pythagorean doctrine of the primacy of number, he depicted the essence of the universe, its Soul, as a numerical sequence containing every canonical form of harmony and proportion discovered in music, geometry and the philosophical study of nature. The World-soul is more than a scale of music, which is but one of its aspects. Throughout his account of how it was constructed Plato clearly implies that it was first and foremost a code of number. From it he derived the natural formulas which, he believed, should be followed by artists and musicians and which he made the basis of his own philosophy.

A riddle which has defeated some of the sharpest minds from classical times onwards is not likely to be solved by any of the methods previously tried. The quest is for a particular number, which Plato referred to as 'the source and origin of motion' and which his follower, Xenocrates, the head of his Academy during the fourth century BC, called 'a self-moving number' and the cause of all natural integers. In the *Timaeus* the Soul which bears this number is said to react to every part of the numerically created universe. That implies that the number of Soul has a wide range of mathematical relations and sums up in itself the basic numbers of creation. No such qualities are found in any of the aggregate numbers so far proposed as the sum of the terms in the World-soul series. The confusion of Plato scholars on this question suggests that some vital clue is missing or has been overlooked in Plato's text, making the problem as it stands insoluble. In that case the only way forward is by a leap of conjecture, by looking directly for a number with the mathematical properties which qualify it to be the number of the World-soul.

The number to be looked for must presumably be within the range of those which most ancient and modern commentators agree in making the aggregate number of Soul. That places it somewhere between A.E. Taylor's 105113 made up of 34 terms and the aggregate of 114695 from 36 terms given in the *Timaeus Lochris*. Then, 36, one of the basic numbers of Platonic arithmetic, is obviously best suited to represent the number of terms in the Soul series, so the aggregate number is most likely to be found in the higher part of the range. It should be a superabundant number, rich in factors, and those factors should include as many as possible of the numbers to which Plato draws attention elsewhere and which are prominent in traditional schemes of cosmology, metrology and music. The ideal number would have factors including:

(*a*) the basic numbers 1–12;
(*b*) the duodecimal numbers constantly referred to by Plato, such as 36, 72, 108, 216, 864, 1296, 1728, etc.;
(*c*) as many as possible of the terms comprising the World-soul;
(*d*) those numbers which are found to have had the greatest significance in ancient number symbolism, such as 3168 and 5040.

There are several numbers within the likely range which fulfil some of these conditions. One of them is the number 108 864 which features strongly in other Platonic schemes, reviewed in previous chapters. It is divisible by all the numbers 1–9 apart from 5 and is the product of 504 × 216. The most outstanding, however, is a number which has also been noticed previously in this book, the number 114 048.

This number agrees with almost all the conditions mentioned above.

(*a*) As 1 140 480 it is divisible by all the numbers 1–12 except 7;

(*b*) 1 140 480 is divisible by the powers of 12 up to 12^4 and by the main duodecimal numbers as exemplified above;

(*c*) the first and last terms in the World-soul series, 384 and 10 368, are factors of 114 048, and so also are one-third of the numbers which the authorities have commonly attributed to the Soul, that is: 384, 432, 576, 648, 864, 1152, 1296, 1728, 2592, 3456, 5184, 10 368;

(*d*) 114 048 is equal to 3168 × 36. If there were meant to be 36 terms in the World-soul, the average value of all the terms is that key number of ancient symbolism, 3168.

The number 504 is not a factor of 114 048, but the two numbers are intimately related in the geometry of a circle. The examples below also show some of the many ways in which the hypothetical 'Soul number' 114 048 relates to the basic numbers 1–10 and 1–12.

1. A circle with circumference of 11 404 800 has a diameter of 3 628 800 which is equal to 5040 × 720 or to 10!, the first ten numbers multiplied together. The area of this circle is equal to 12! × 21 600.
2. If 11 404 800 represents the diameter of a circle, its circumference ($\pi = 864/275$) is exactly 12^7. The side of a square inscribed within a circle of diameter 114 048 is equal to twice 8!.

Many other examples could be given of the connection between 114 048 and the numbers 1–10, the terms in the World-soul and the other numbers which are characteristic of Pythagorean arithmetic. Some of the functions of 114 048 in various departments of ancient science have already been noticed (see Index of Numbers): for instance, there are 11 404 800 feet in the moon's diameter, and 114 048 000 is the number of Egyptian feet in the earth's circumference. In the canonical scheme of New Jerusalem or Magnesia the total area minus the area of the central city is also 114 048 000.

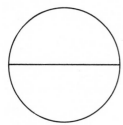

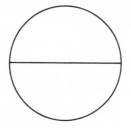

Earth measured in units of 10 Egyptian feet
(11.52 ft.):
diameter   = 10!
circumference = 114048

Moon measured in ft.:
diameter   = 11 404 800
circumference = 7!

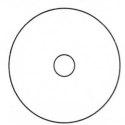

The New Jerusalem:
diameter of inner circle = 1080 miles
                                     = 11 404 800 ft.
diameter of outer circle = 6120 miles
area of outer minus
inner circle   = 114 048 000
                                    sq. miles.

side = 1 140 480 ft.
area = 10×12^6
        acres
     = $6 \times 12!$ NJ
        units

Figure 64. The number 114 048, its intimacy with the numbers 1 to 12 and its prominence in the dimensions of earth and moon and the New Jerusalem diagram.

From the evidence which has been accumulated in earlier chapters of the derivation of Platonic number from the ancient tradition, there seems little doubt that Plato would have recognized the outstanding qualities of the number 114 048 which befit it above all others to be called the number of the World-soul. The traditional description of that number, as the cause of motion and the source of all natural integers, applies perfectly to 114 048 which, as shown above, can be said to engender the first ten or twelve basic numbers as well as being related to all the main symbolic numbers in the ancient system.

There remains, however, the problem of reconciling this number with the aggregate number of the terms in the World-soul. It is smaller by a difference of 647 than 114695, which is the aggregate of the 36 terms given in the *Timaeus Lochris*, and though the difference can virtually be eliminated by dropping one of those terms, 648, there is no apparent justification for doing so. Perhaps Plato himself was unable to find an exact equation between the total value of the terms in the Soul and the number which ideally should represent it – which would explain the ambiguities in his instructions. It can therefore not be claimed that the identification of 114048 as the number of Soul is proven absolutely. It remains a hypothesis, the usefulness of which will be determined by whether or not it proves to be of assistance in future studies of symbolic number.

The hypothesis that 114048 was intended to be the aggregate number of the World-soul leads naturally to the unfolding of a geometrical diagram illustrating the Creator's subsequent dealings with the Soul material. Having used up all the mixture by removing the various portions of it, he was left with a strip of material. This was of a certain length and of a certain breadth and also, presumably, of a certain thickness, although, since it was later formed into two equal circles revolving one inside the other, the thickness must have been negligible. In that case, 114048 being the area of the strip, the question concerns its length and breadth.

Figure 65 illustrates the answer which seems most likely. The original, undivided strip has been taken to measure 144 in breadth and 792 in length, the two dimensions multiplied together making 114048. The ratio between the breadth and the length, 2:11, is the same as between the sum of the original seven terms in the World-soul (20736) and the supposed aggregate number of all the terms (114048).

When cut in two lengthways down the centre, the strip becomes two strips, each measuring 72×792. These are placed one over the other to form a cross. The square at the centre where they overlap has an area of 72^2 or 5184. Thus the visible area remaining is $114048 - 5184 = 108864$, and 108864 has previously been identified as an important number in Platonic arithmetic: in the plan of Magnesia 108864 NJ units is the total amount of territory to be shared among the 5040 citizens. The square containing the crossed strips of Soul material is numerically the same as the square form of Magnesia (figure 48).

The two strips are now bent into circles and are joined together again opposite their first junction. The outer circle, the Same, is the dominant and

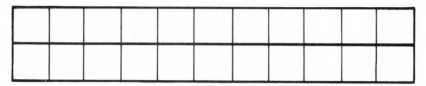

Figure 65. The material of the World-soul is shaped into a strip made up of 22 squares, each of area 5184 or 72^2, the total area of the strip being 114048 and its dimensions 144 × 792. Four of the squares are taken up by the original seven numbers of the World-soul amounting to 20736.

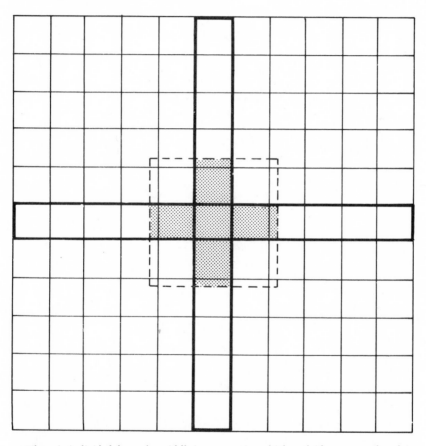

The strip is divided down the middle into two strips which are laid one across the other. Their total superficial area is thus reduced to 108864. When the two strips are bent upwards into circles and joined to each other at the top, they appear from above to retain the form of a cross (shaded) and to occupy an area equal to six of the original World-soul squares. The side of the square containing the apparent cross measures 252.

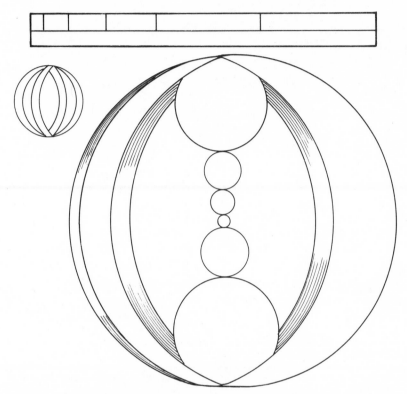

Figure 66. The strip forming the 'circle of the Other' is divided into seven parts in the proportions 1, 2, 3, 4, 8, 9, 27 by making six cuts down the lines marked above. The longer strip resumes its function as the circle of the Other, while the other six are bent into circles, fitting together within the first.

Below, the six smaller strips, bent into circles, are represented by six globes contained within the circle of the Same and the (reduced to half-width) circle of the Other.

perfect one, so, even if the strips were modelled for Plato's demonstrations out of thin parchment, the inner circle, the Other, must be conceived of as slightly distorted, being necessarily flatter at its poles than at its equator. This makes it an image of the earth. Since the length of each strip of Soul material is 792, the diameter is 252.

The circle of the Other is then split into seven unequal circles by six cuts, so that the seven circles are in the proportions 1, 2, 3, 4, 8, 9, 27. How this is done is shown in figure 66. The circle of the Other, though halved in width, retains its former length of circumference. The six strips cut off it are bent into circles whose combined diameters are of the same length as the diameter of the circle of the Other. They will therefore fit together along its axis as in figure 66.

These six circles are made to revolve in the various ways Plato indicates within the circle of the Other. His description of their motions and their relationships to the planetary orbits is so obscure and complicated that no one from Aristotle onwards has claimed to be much the wiser for studying it. Plato himself said (*Timaeus* 40D) that the astronomical system he was referring to could not be made clear without displaying its model. The model he had before him was probably a metal armillary sphere which he used for instructing the pupils in his Academy on the mechanics of celestial motion.

The conclusion reached from these studies of Plato's mathematical allegories is that they were based on a traditional codification of number itself, developed from the numbers 1 to 12, summed up in the numbers 5040 and 7920 and culminating in 11!, or, more completely, in the number $479\,001\,600 = 12!$. The nodal points in this structure, the points where the different orders of number mingle or intersect, are occupied by those numbers which occur throughout the ancient codes of science, such as the factors and multiples of 5184, 108 864, 114 048 and others here listed in the Index of Numbers.

Plato's evident intention was to identify number as the archetype of creation and to draw attention to those particular numbers which constitute its core and which, in various combinations, generate the entire field of number. His purpose was to provide for the use of rulers and reformers an objective standard, thus allowing statecraft to become a science, firmly based on those numbers and proportions which constitute the essence of the universe.

5 Symbolic number

THE FIRST CHARACTERISTIC of the Heavenly Jerusalem is completeness. As a symbol of the entire universe it should display in its dimensions the number representing every god, power or tendency in nature. Some of the principal numbers in the scheme have here been identified; among them are 3168, 1080, 1224 and 1746, and there is reference also to 864 and 666. All these numbers were applied by gematria to important names and phrases in the sacred texts of early Christianity, which provides clues to their symbolism, and further inferences are derived from their astronomical and other associations. They exemplify the manifold significance which the ancients discovered in certain numbers and which caused them to make use of those numbers for codifying their knowledge of the world.

864, the foundation number

The strong, solar character of the number 864 is consistently brought out by the phrases associated with it by gematria and by its astronomical functions. It is a number of four-square order and firm foundation.

An image of 864 is the rock at the centre of the earth holding down the waters of the abyss, which is symbolized by the foundation or corner stone of a temple and by its altar. Thus the corner of the earth on which the angel stood in Revelation (7,1) is γωνια 864, and 864 is the number of an altar, θυσιαστηριον. By the rules of gematria the letters στ in that word can be given the combined value of 6, making the number of an altar 864, or they can be counted as 500, in which case the number of το θυσιαστηριον, the altar, is 1728 or twice 864.

The diameter of the sun measures 864 000 miles, and the character of Jerusalem as a world-centre is expressed by its gematria, 'Ιερουσαλημ, 864. The holy of holies within its temple is ἁγιων, an anagram of the word for corner stone and similarly occupying a central point. Its number, 864, is shared by the temple of the gods, θεων. Also amounting to 864 are the combined values of the names 'Αθηνα, 69, and 'Ηφαιστος, 795, Athena and Hephaistos, the divine founders of Athens, where the Acropolis corresponds to the sacred rock at Jerusalem.

The cube of the New Jerusalem, measuring 12 000 or on a reduced scale 12 furlongs on every side, has a superficial area over its six sides of 864 square furlongs or 8640 acres, and its volume in cubic furlongs is $864 \times 2 = 1728$. St Augustine in Book xv of *The City of God* compares the cube with Noah's Ark, which 'being all of square wood signifies the unmoved constancy of the saints [ἁγιων = 864]; for cast a cube or square body which way you will it will ever stand firm'. A geometer's image of the Church built on the central rock is a square of area 86 400 and side measuring about 294, the value of the word ἐκκλησια, church, making the measure round its four sides 1176 = υἱος μονογενης, only-begotten Son. When the waters of the abyss rise up and overwhelm the rock, together with the Church and all earthly structures, the symbolism of the rock is transferred to the ark, which carries the sacred measures and other elements of culture from one civilization to the next. As the man who discovered firm ground for the renewal of traditional culture, Pythagoras, Πυθαγορας, had the appro-priate number, 864.

The solar associations of this number begin with the 864 000-mile diameter of the sun, and traditional measurement of time, governed principally by the sun, feature the number 864 on all scales, from the period of 8640 million years or one day and one night of Brahma to the 86 400 seconds in a 24-hour day. A name for the divine ruler of the 365 days in a solar year was Abraxas, Ἀβραξας, 365, and his seat, the throne of Abraxas, θρονος Ἀβραξας, has the number 864.

In the language of symbolic number 864 clearly pertains to a centre of radiant energy, the sun in the solar system, Jerusalem on earth, the inner sanctuary of the temple, the altar within it and the corner stone on which the whole sacred edifice is founded. Its characteristic form is the cube of New Jerusalem with 12 units to a side and volume of twice 864 cubic units. That is the foundation rock, placed at a point of union between heaven and earth, where the rays of the sun penetrate the realm of the earth spirit. Its rectilinear symmetry identifies it as an artifact, a product of reason and one of the two elements which comprise the ideal cosmology symbolized by the squared circle. The other, circular element, which is required to balance the four-square, man-made city, is the heavenly form, the City of God. Its number (ἡ πολις θεου = 882) combines with 864 to make the sum 1746, the Number of Fusion.

3168, the perimeter of the New Jerusalem

The paramount importance of 3168 in the traditional canon of number is indicated by its leading position in the New Jerusalem scheme, and also by its adoption by the founders of Christianity as the number of their first sacred name, Lord Jesus Christ. There is mystery about the origins and meanings of the three terms in that name, the first of which, *Kyrios* or Lord, is the title given by astrologers to the dominant influence of an age. Added to Jesus Christ (᾿Ιησους Χριστος, 2368), the title Lord, Κυριος, 800, completes the number appropriate to Lord Jesus Christ, 3168.

The number 3168 is superabundant, the Pythagorean term for a number which is exceeded by the sum of its factors. The sum of all the numbers which divide into 3168 is 6660, connecting the number of Lord Jesus Christ with that of the Beast in Revelation. This subject is explored later in the section on the number 666.

Another phrase with the number 3168, in St Paul's Epistle to the Romans (3, 22), is πιστις ᾿Ιησου Χριστου, Faith of Jesus Christ, and 3168 is also the value of ᾿Ιησους Χριστος κοσμου, the cosmic Jesus Christ. The pre-Christian reference of this number seems to have been to the twelve gods of the zodiac. In the centre of the marketplace in Athens was the *omphalos* pillar, from which all distances were measured and where the sacred paths converged. The inscription carved on it, οἱ δωδεκα θεοι, the Twelve Gods, has the value 1008 which is the diameter of a circle with circumference 3168, and 3168 is the number of τεμενος δωδεκα θεων κοσμου, sanctuary of the twelve gods of the cosmos. One unit in excess of 3168 numbers the phrase ναος των δωδεκα θεων, temple of the twelve gods. Thus the inscription on the pillar to the twelve gods can be seen as a geometric allusion to the sacred area around it, enclosed by a ring of twelve shrines to the powers of the universe, as in the New Jerusalem diagram.

The dimensions of this circle dedicated to the twelve gods were readily adapted by gematria to Christian terminology, for

1008 = οἱ δωδεκα ἁγιοι, the twelve saints
3168 = τα ἱερα των ἀποστολων, the shrines of the Apostles

These phrases seem particularly appropriate to the foundation plan at Glastonbury, where the perimeter of the square of 12 hides in which the 12 founding saints placed their ring of cells is 31 680 ft.

In the New Jerusalem diagram 31 680 measures both the perimeter of the square and the circumference of the circle, and it permeates the entire ancient system of number and measure. It is the natural partner of 5040, the number on which Plato proposed to found Magnesia, being related to it as the circumference to the radius of a circle. An example of the collaboration of these two numbers in ancient metrology is given on page xx. Together they provide the chief symbols of the traditional canon of number which forms the inner structure of number itself.

Some of the numerical qualities of 3168 and its appearances in ancient cosmology are summarized below.

The circle with circumference 31 680 and radius 5040 consists of two semicircles each with an area equal to 11!.

The perimeter of the square New Jerusalem is 4 × 12 furlongs or 31 680 ft.

The perimeter of the square 12 hides of Glastonbury is 31 680 ft.

The mean circumference of the Stonehenge sarsen circle is 316.8 ft. or a hundredth part of 6 miles.

A square containing the circle of the earth, average diameter 7920 miles, has a perimeter of 31 680 miles, and if the moon, diameter 2160 miles, is drawn tangent to the earth, a circle struck from the centre of the earth circle to pass through the centre of the moon has a circumference of 31 680 miles.

In his *Natural History*, at the end of the second book, Pliny implies that the world-circumference is 3 168 000 miles.

The number 3168 has been preserved in our traditional system of metrology from at least as early as the building of Stonehenge. For example,

31 680 × 2 inches = one mile
31 680 ft.   = 6 miles
31 680 furlongs   = 3960 miles = mean radius of the earth.

The longer Greek furlong or stade of 625 Greek feet, which appears in the dimensions of the Parthenon and Stonehenge, is equal to 633.6 or twice 316.8 ft. This Greek furlong and the English furlong of 660 ft. come together in the acre. For example, a rectangle with sides measuring one English by five Greek furlongs (660 × 3168 ft.) contains 48 acres exactly, and the 1440 acres of the New Jerusalem square are also contained by a rectangle of 3 English furlongs by 5 Greek furlongs.

The general character of the number 3168, as conveyed by its position in ancient cosmological diagrams and the phrases associated with it through gematria, is that it represents the spirit which passes through and encircles the universe, Plato's World-soul. The Christian term for this spirit, developed from the number 3168, was Lord Jesus Christ.

1224, the number of Paradise, and the 153 fishes in the net

The total width of the New Jerusalem diagram is equal to the diameter of the earth plus two diameters of the moon, or 12240 miles. Thus the number 1224 refers to the cosmological city, image of God's creation. This interpretion accords well with the gematria of 1224.

1224 = ὁ κυριος ὁ θεος, the Lord God
    = κτισις θεου, God's creation
    = κυκλος θεου, divine circle
    = ἡ φυτεια, the plantation (Matthew 5,13), an early Christian synonym of paradise

Similar meanings are found in the gematria of the numbers on either side of 1224.

1223 = ἡ ὁδος παραδεισου, the way of paradise
1225 = Ὁ παραδεισος θεου, God's paradise (Revelation 2, 7)
    = ἐγω εἰμι ἡ ὁδος, I am the way (John 14, 6)
    = ἀρχηγετης, founder of the city
    = ἡ δικαιοσυνη θεου, the righteousness of God (Romans 3, 22)
    = ἐν ὁλον ὁλων, the entirety of creation (literally, One Whole of Wholes), Plato's phrase in *Timaeus* for the unique sphere of the cosmos embracing all its parts.

The number 1224 links the foot and the megalithic yard (the unit of 2.72 feet deduced from measurement of megalithic monuments). There are 1224 ft. in 450 megalithic yards and 1224 megalithic yards in 3330 ft. The number 1225 is triangular, being equal to the sum of the numbers from 1 to 49.

Several of the incidents and parables in the New Testament story of Jesus are known to have been adapted from earlier writings, and some of these have hidden meanings which the Christian gnostics interpreted by the same cabalistic methods as the Jews apply to the exegesis of their own scriptures. Most obviously numerical is the tale of the miraculous catch of 153 fishes which occurs in the last chapter of St John's Gospel. Why there should have been exactly 153 fishes in the net which the Apostles cast into the Sea of Tiberias is a question which has puzzled commentators from early Christian times. A clue which previous writers have noticed is that two of the key words in the story, ἰχθυες, fishes, and το δικτυον, the net, each have the value by gematria of 1224, and 1224 is 8 times 153.

Figure 67. Seven circles, representing six disciples with Simon Peter at their centre, pack into the circular boat of diameter 1224.

Following up this clue we are led on to reconstruct the figure of sacred geometry which must originally have accompanied the story of the 153 fishes. It develops in three stages, reflecting the order of events in John 21.

After the Crucifixion, Simon Peter went fishing from a boat in the sea of Tiberias, taking with him six of the other disciples. They fished all night but caught nothing.

In the morning they saw the risen Jesus on the shore, but failed to recognize him. He called out that they should cast their net on the right side of the boat. Having done so, they were unable to draw it out for the multitude of fishes in it.

John then recognized Jesus and told Simon Peter, who put on his fisher's coat and jumped into the sea. The other disciples followed him in the boat to the shore, which was about 200 cubits away, dragging the net with the fishes. When Simon Peter drew it to land it was found to be 'full of great fishes, an hundred and fifty and three: and for all there were so many, yet was not the net broken'.

The number of Σιμων ὁ Πετρος, Simon Peter, is 1925, so Peter can be represented by a circle with circumference 1925 and diameter 612½ or 612. This is appropriate because 612 is the number of ὁ ποιμην ἀγαθος, the Good Shepherd, and that is the title which Simon Peter inherits when, following the incident of the 153 fishes, he is told three times by Jesus, 'Feed my sheep.'

Six more circles of the same dimensions are drawn for the six other

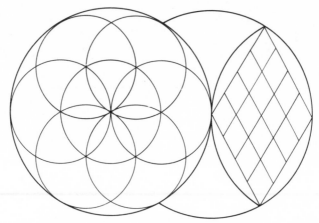

Figure 68. A net formed by a rhombus within a Vesica Piscis is cast on the right side of the boat, catching 153 fishes.

disciples, and the seven are packed together in the most economical way (figure 67) and placed inside the circular boat, like the coracle of the Celtic saints, the diameter of which is 1224.

The disciples are told to cast their net on the right side of the boat. This is done geometrically by placing the compass point on the circumference of the circular boat and drawing an arc of another circle with the same radius, containing a Vesica Piscis (vessel of the fish). The rhombus within it is divided up into sixteen smaller diamond shapes (figure 68). Its width being 612, each of its sixteen divisions has a width of 153. They represent sixteen small fishes making up a greater seventeenth, and here again the number 153 is brought out, for 153 is the sum of the numbers from 1 to 17. The measure round the four sides of the greater rhombus-fish is 2448 or 1224 + 1224 or τo δικτυον, the net, plus ἰχθύες, fishes. Thus the net full of 153 fishes is illustrated in number and geometry.

'Now when Simon Peter heard it was the Lord, he girt his fisher's coat unto him (for he was naked), and did cast himself into the sea.'

This incident is depicted in figure 69, where another Vesica is formed on the left side of the boat, and the circle representing Simon Peter is taken from the centre of the boat and placed within it. The width of the Vesica is 612, so its height is 1060, and that is the number of ἡ ἐπενδύτης, the fisher's coat. Simon Peter, the naked circle, is now clad in the Vesica of his coat. He is between the boat and the land, which is said to be about 200 cubits distant. The Greek cubit was slightly over 1.52 ft. Taking it as 1.53 ft. makes 200

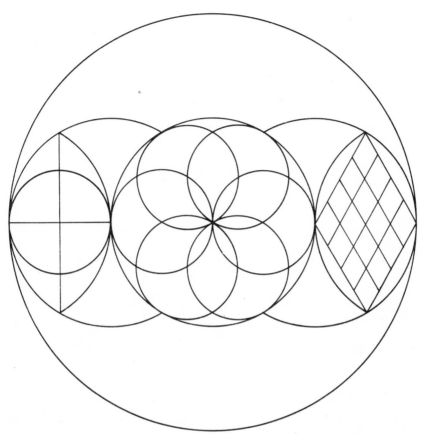

Figure 69. Simon Peter, wearing his fisher's coat, jumps into the sea to the left of the boat and tows it to land. The circle round the entire figure represents the Sea of Tiberias, diameter 2448.

cubits equivalent to 306 ft., and 306 is the measure between the centre of the Simon Peter circle and the circle of land surrounding the whole figure. The diameter of this enclosing circle is 2448, closely corresponding to the number of the name which St John (6, 1) applies to the scene of the miraculous catch, 'the sea of Galilee which is the sea of Tiberias', θαλασσα Γαλιλαιας της Τιβεριαδος, 2446.

St Augustine and other early commentators take the 153 fishes as a symbol of the number of those on earth who are destined for salvation. The net signifies the whole world, and its number, 1224, identifies το δικτυον with the phrases listed above, such as God's creation, which have by gematria the same numerical value. Further insight into the meaning behind the tale of the 153 fishes may be obtained through studying the pattern of its geometric accompaniment. The three intersecting circles produce the central diagram of Hebrew mysticism, the Tree of Life, ξυλον ζωης, 1625, corresponding to the Pythagorean Tetractys, τετρακτυς, 1626. In the final episode of the story the disciples come to land, where Jesus awaits them with fish laid on a fire and bread. Fish, fire and bread represent three worlds united in the cabalists' Tree, fish being a symbol of the mercurial spirit which mediates between the elemental world above and the material below.

It is a traditional practice among teachers of esoteric philosophy to set forth their doctrines in the guise of simple parables which amuse children, enrich popular mythology and, for those who understand the science of interpreting them, illustrate various cosmological processes. The themes which are adopted by hagiographers and composers of sacred legends are those which occur spontaneously in different times and cultures and can therefore be called archetypal. Thus the founders of Christianity took certain episodes in universal folklore and made Jesus their central figure. In the tale of the 153 fishes he plays the part of the shamanic man of miracles whose traditional functions include bringing good luck to hunters or fishermen. By interpolation of names and numbers this story was made to reflect the construction of a geometric diagram with cosmological significance, by reference to which the gnostic masters were able to demonstrate to initiates the basic truth behind the Christian legend.

1080, the lunar number

Of the two forces or tendencies which, through their interactions, create the apparent universe, that which is called *yin* or receptive – the lunar, earthly, inspirational as opposed to the solar, cosmic, rational side of nature – is subsumed under the number 1080.

The philosophy of number is subtle and easily misunderstood, which is why we do not in the first instance define 1080 and its polar opposite, 666, as the numbers of the female and male. It is appropriate to do so in a secondary sense, but the division of species into two sexes is not the primary cause of universal duality, merely one of its effects. The Chinese terms *yin* and *yang* are now widely used, their advantage being that they are abstract and unemotional, as also are numbers. A number on its own has no inherent meaning; its symbolism arises purely from its relationships to other numbers.

It has no moral qualities, nor can it be an absolute symbol of any created thing. The absolute *yin* or negatively charged element and its opposite, the positively charged *yang*, are creatures of concept alone, unmanifest. Yet everything in the universe can be seen as representing one or the other of the two forces in relation to all other things which are comparable to it. The *yin* number 1080 can thus be applied to:

the horizontal in contrast to the vertical	
below	above
dark	light
subtlety	rigour
the receptive	the generative
contraction	expansion

and so on. As long as one keeps to abstract principles the matter is clear enough. But the further one descends into the realm of the material the more ambiguities arise. Human functions can be seen in terms of polarity, the ruler being positive in relation to his subjects, but in the ordinary course of life people's roles and their conceptual polarities are constantly being reversed; and on the lowest level one meets the same problem as did Socrates over his theory of archetypes: that one can hardly suppose that there is an ideal form of, say, the dirt on the street, nor can its place in the world-order be described through number.

Traditional philosophy is not however concerned with such random effects as mud, nor with any of the actual manifestations of nature. The philosopher's study is of eternal reality, and this is not an attribute of the apparent universe which, having been born, is therefore inconstant, mortal and a mere reflection of the causes behind it. In the beginning, it is said, the universe reposed as in a womb; heaven and earth embraced, the opposites were undifferentiated and the laws which guided its post-natal movements were inherent but yet inactive. By prising heaven and earth apart or, as told in Genesis, by separating the darkness from the light, the Creator set his work in motion, producing tension between its two primary components and thus initiating the process by which the solid universe was formed, as a pearl is formed between the two halves of an oyster. Sacred and beautiful though it is as a product of divine nature, the pearl is of less interest to the philosopher than the creative laws that produced it; and the laws which fashioned the universe are said to be the laws of number, established by the Creator's original act of thought. Every numerical formula obviously involves two or more numbers, which in their various interactions can be seen as positive or negative. Thus the traditional image of the world is a magnetic field, a paradoxical structure in which the particles are illusory but their relationships real and constant.

The philosopher who acquires this view of things cultivates the habit of distinguishing and qualifying the opposite elements in any situation, and applying his knowledge of proportion to locating their point of balance. To achieve this he may take sides and adopt causes, but he is incapable of fanaticism or moral self-righteousness, seeing in everything the operations of two rival but interdependent forces, neither of which can entirely vanquish the other nor long prevail over it without producing reaction.

The number 1080, the *yin* term in the cabalists' equation $666 + 1080 = 1746$, is identified by all its symbolism with the moon, the sublunary world, the waters below that are drawn by the moon, the lunar influence on the earth's vital currents, the periods of the female, the unconscious, intuitive part of the mind and the spirit that moves oracles. It also relates to the measurement of time and the heavenly bodies. This arises because the moon is said to be the first source and standard of measure. There is similarity in the Greek words for measure, mother, moon and month (metron, meeteer, meenee, meen), their common initial prompting the suggestion (sanctioned by Plato's etymological methods in *Cratylus*) that the characteristic sound of the number 1080 is the letter 'm'.

As a lunar number, 1080 measures the radius of the moon in miles, and the corresponding principle to the moon in the world of minerals, silver, has an atomic weight of 108. The geometric image of that same principle, the pentagon, exhibits the angle of 108° between its sides.

In chronology 1080 is traditionally the number of breaths one takes in an hour, and the Jews therefore divide the hour into 1080 minims or *chalakim*. Twice 1080 years is the length of a 'month' in the Great Year, 2160 years being the period in which the sun progresses through one sign of the zodiac, and 108 000 years is the duration of a season in the Hindu Kali Yuga of 432 000 years. According to Heraclitus, civilization is destroyed every 10 800 years. Modern climatologists give approximately the same period as the interval between successive ice ages.

The old astronomers' use of sexagesimal number makes 10 800 the number of minutes in the semi-circumference of the earth or in any semicircle, and it is also the number of seconds in three degrees of a circle or three hours of time. As a symbolic number, 1080 is given by Hipparchus as the number of stars of first magnitude brightness, and Galileo wrote that the sun's diameter contains the diameter of a sixth-magnitude star 2160 or twice 1080 times.

108 is an important natural ratio in astronomy, being the mean distance between the earth and the sun, measured by the sun's diameter. The same distance is equal to 43 200 or 4 × 10 800 diameters of the moon.

108 This number is widely referred to in religious symbolism. There are, for example, 108 beads in the Hindu or Buddhist rosary, 10 800 stanzas in the *Rigveda*, each of 40 syllables, and 10 800 bricks in the Indian fire altar. In the Norse Eddas there are said to be 540 doors to Valhalla, where the shades of *5 × 108* past heroes enjoy perpetual fighting and feasting. Each door being double and framed by two pillars makes the pillars round Valhalla number 1080.

The ratios of ancient metrology are also based on the number system to which 1080 belongs. The Roman half-pace of 1.216 512 ft. divides exactly 108 000 000 times into the earth's mean circumference, and 1080 square megalithic yards are equal to 888 square yards in English measure. By gematria the number of the phrase $\tau o\ \theta\epsilon\iota o\nu\ \mu\epsilon\tau\rho o\nu$, the Divine Measure, is one less than 1080.

The symbolic character of 1080 is clearly defined by its gematria, which shows it to have had the same meaning to the early Christians as to the pagan philosophers. As the number of the *yin* in nature it is identified with the third component in the Trinity, the Holy Spirit, which is also the Spirit of the Earth. These two phrases in Greek have not only the same number but the same letters.

> 1080 = $\tau o\ \dot{\alpha}\gamma\iota o\nu\ \pi\nu\epsilon\upsilon\mu\alpha$, the Holy Spirit
>   = $\tau o\ \gamma\alpha\iota o\nu\ \pi\nu\epsilon\upsilon\mu\alpha$, the Earth Spirit

and

> 1080 = $\dot{o}\ \theta\epsilon o s\ \pi\alpha\rho\theta\epsilon\nu o s\ \gamma\eta s$, the virgin god of earth
>   = $\pi\eta\gamma\eta\ \sigma o\phi\iota\alpha s$, fountain of wisdom
>   = $\dot{\eta}\ T\alpha\rho\tau\alpha\rho o s$, Tartaros, the nether world

1081 is the number of $\dot{\alpha}\beta\upsilon\sigma\sigma o s$, the Abyss, and of the regional spirits called $\Sigma\alpha\tau\upsilon\rho o\iota$, Satyrs. Other adjacent numbers produce phrases which are consistent in their meanings with those above, such as 1082 which is $\dot{\eta}$ $\pi\rho o\phi\eta\tau\epsilon\iota\alpha$, prophecy, $o\dot{\upsilon}\rho o\beta o\rho o s$, the ouroboros serpent, and $\dot{\eta}\ \pi\eta\gamma\eta$ $\dot{\upsilon}\delta\alpha\tau o s$, the 'well of water springing up into everlasting life' (John 4, 14). 1079 is the number of $\dot{o}\ \chi\theta o\nu\iota o s$, the god of the underworld.

The Earth Spirit, 1080, corresponds to the Chinese *ch'i* or life-breath of nature, which accumulates in folds and cavities of the earth, giving to certain places that peaceful, other-worldly atmosphere which marks them as natural centres of healing and oracles. At every spot the local character of the Earth Spirit is conditioned by topography and also by the sun and other cosmic influences. These representatives of the number 666 give to those of 1080 their beneficial qualities, and if there is no union between them the Earth Spirit becomes sour and virulent. In deep subterranean caverns, where the light of the sun is unknown, dwell those monstrous, atavistic creatures of the underworld, the demons that watch over buried treasure and the phantom forms which haunt the night-side of nature and lurk within the dark recesses of the mind. Initiates of the ancient Mysteries, keeping vigil in chambers beneath the earth, had personal experience of these dread images, and in some cases it overwhelmed their sanity. Those who survived the ordeal were reborn and, having outfaced the terrors of darkness and death, had no further fears in life. Following their union with the Earth Spirit they were called Bridegrooms (νυμφιοι, 1080).

Thus 1080 is the number of the Earth Spirit as the source of Universal Harmony (ἡ ἁρμονια κοσμου = 1080), Prophecy (1082) and Wisdom (πηγη σοφιας, 1080), and it also stands for Tartaros (1080) and the Abyss (1081). It is the number of magic, imagination and madness and, above all, of that Mystery or principle of equivocation, which lies at the heart of things and is not to be comprehended by any system of morality or rationalism.

The founders of Christianity, who named the third part of their triune deity το ἁγιον πνευμα which has the number 1080, evidently regarded the Holy Spirit in a very different light from that in which it was later held by the Church. The gnostics invoked the powers of 1080 through the practices they adopted from previous religions, cultivating oracles and mediums and thus acknowledging the eternal feminine principle which the Church Fathers tended to identify with the satanic serpent. Following the suppression of gnosticism, the Holy Spirit lost many of its traditional attributes, the *yin* element which it represented being out of favour with the authorities. Necessity, however, compelled its restoration, and it re-entered Christianity in the image of the Madonna and Child (Mary, Μαριαμ, 192 + Jesus, 'Ιμσους, 888 = 1080), whose popular cult reactivated the old shrines of the Earth Spirit and permitted once more the expression of reverence for nature. The Church was relentless in opposing all other associations of the number 1080, such as the earth-serpent, the hermetic spirit, the necromantic oracle and the alchemical science which gave that number its meaning. Even now there is prejudice against its study from those who associate it exclusively with sorcery and the black arts.

An earlier enemy of the old science was Jewish monotheism. In Chapter 8 of Ezekiel the prophet describes with horror the solar and underground cults which flourished at the Temple. In its inner court he saw 'about five and twenty men, with their backs towards the temple of the Lord, and their faces toward the east; and they worshipped the sun toward the east'. In a vision he was shown a hole in the wall of the Temple. Digging into it he uncovered a door, leading to an underground chamber. A dreadful sight was revealed within, of the monstrous progeny of inner earth, 'every form of creeping things, and abominable beasts, and all the idols of the house of Israel, pourtrayed upon the wall round about'. Worshipping these things were 'seventy men of the ancients of the house of Israel'. A voice spoke to the prophet: 'Son of man, hast thou seen what the ancients of the house of Israel do in the dark, every man in the chambers of his imagery? for they say, The Lord seeth us not . . .'

In this strange story Ezekiel appears to denounce both the underground cult which took place in a chamber beneath the Temple and also the unclean fantasies that sullied the minds of the priests. He evidently understood the Temple as an image of human mentality with its solar, rational element above ground and its lunar, subconscious side buried beneath. Archaeology has shown that ancient temples were indeed built in this way, their upper parts orientated towards the heavenly bodies and in symbolic accordance with the cosmos, while below in the domain of the earth-goddess were caves and catacombs of initiation. Examples of these occur beneath the temples at Jerusalem, Hieropolis and elsewhere, and they are referred to in legends as the haunts of monsters, such as the Minotaur in the Cretan labyrinth.

Herodotus, who visited the great labyrinth in Egypt but was not allowed to enter it, reported that it contained 1500 chambers below ground, reflecting a similar number above, and that in its depths lived sacred crocodiles. In modern folklore the crocodile occurs spontaneously as a denizen of the underworld. From time to time there are rumours that the sewers below New York and other modern cities harbour colonies of crocodiles, and, strangely enough, actual crocodiles are sometimes found in them. One recent example, of a crocodile taken from the sewers of Paris, was reported in *The Times*, 9 March 1984: 'Municipal workers found it and called in firemen to muzzle it before it could be taken to a zoo. How it entered the sewer system remains a mystery.' Mystery indeed! — and a hint that behind rumours of crocodiles and their occasional appearances in city drains is a mystery more profound. In the study of ancient symbolism one finds many instances of its themes recurring in fairy-tales, folklore and popular journalism today. This clearly is evidence of the eternal and archetypal nature of such themes. And when they take physical shape, as the oracular crocodiles in pools beneath Egyptian temples and the errant creatures in modern urban sewers, one is led to suspect interplay between archetypal images of the mind and corresponding forms in the world of phenomena. One is led also to wonder at the psychological insight of the ancient philosophers which enabled them to represent the human mind and the nature of the universe together in one scheme, the cosmic temple.

By drawing its main distinction between what it supposes to be the real and the imaginary, ascribing the one to science and the other to poetry, modern philosophy has deprived itself of conceptions such as the Earth Spirit which are creatures equally of nature and the mind. The ancients did not categorize in that way. They took the world and human nature as they found them, recognized the two opposite tendencies that comprise them and paid as much respect to the dark, irrational side of things as to the opposite principle which appeals to the solar intellect. They were not concerned with secondary questions such as morals, which they regarded as mere *mores*, sets of customs which are more or less appropriate to different types of societies. Thus they avoided the pernicious habit of investing the eternal elements of nature, and the symbolic numbers that describe them, with moral qualities and adjudging them good or evil, real or imaginary. Being accustomed to the modern dogmatic styles of religion and science, we find much that is puzzling in ancient philosophy and the science that developed from it. It is easy enough to understand the polarization of nature in a metaphysical sense, but in studying the practical use to which the ancients put their philosophy we are in unfamiliar territory, which can not be explored with the intellectual equipment used today. Modern positivist philosophy, limiting its inquiries to phenomena which conform to known physical laws, thereby cuts itself off from whole areas of reality, recognized by the ancients. Thus we have no modern term for describing that important aspect of the world of experience, which has correspondences throughout the different orders of nature, such as the moon, the metal silver, the waters and energies of the earth, the nether regions and the spirit of prophecy, and was known in the language of ancient science by the number 1080.

The modern style of philosophy differs from the traditional in accepting the reality of the palpable universe and encouraging science to investigate material phenomena which, according to the traditional way of thinking, are mere illusions. For many years it has had its way but, to the dismay of its hight priests, modern physicists have turned against it, and the basis of materialism, the notion that matter actually exists, is no longer scientifically tenable. Two interests, of scientific truth and of the survival of the earth as a living planet, bar the way to further development of positivist philosophy and demand that we adopt a more appropriate way of viewing the world. This has long been recognized by those eminent physicists who, having contemplated the philosophical implications of modern research, find themselves in the spiritual presence of old Pythagoras with his dictum, All is Number.

666, the solar number, and the number of the Beast

As 1080 is to the moon and the realm of imagination and mystery, so is 666 to the energy of the sun and the principle of reason, will and authority. It represents the positively charged nucleus of the atom, the might and glory of the emperor and the intellect in the mind of the philosopher. 666 is the generative power of the male, the call to action, the electric impulse which regulates the molecular field and gives form and order to chaos. It is the active, inventive, fertilizing current in nature, the material as opposed to the spiritual side of things.

In the New Jerusalem diagram the number 666 has its due place as the area in square megalithic yards of the circle corresponding to the bluestone circle at Stonehenge with diameter of 79.20 ft.

The other circle in the diagram, diameter 100.8 ft., has an area of 1080 square megalithic yards. The significance of these two numbers, 666 and 1080, occuring together in the New Jerusalem scheme has been commented on earlier, and so has the connection between 6660 and the characteristic New Jerusalem number 3168, which is that the sum of all the numbers which divide into 3168 is 6660.

The number 666 is the most notorious of all symbolic numbers as being that which St John ascribed to the Beast in the last verse of Revelation 13: 'Here is wisdom, let him that hath understanding count the number of the beast: for it is the number of a man; and his number is six hundred threescore and six.'

The interpretation of these words is given in the latter part of this chapter, shedding clear light on St John's gnostic attitude to the Roman Church. For proper understanding of the matter, preliminary study of the number 666 is needed, always bearing in mind that symbolic numbers stand apart from human ideas of good and evil and are without moral connotations. 666 represents the principle of authority, irrespective of whether it be exercised by a wise ruler or a cruel tyrant. Its sinister reputation has been augmented by the translation of the word θηριον (*thērion*), which in Greek means a wild animal, as the Beast. There is nothing inherently beastly or evil about the number 666 although, like all symbolic numbers and the tendencies they symbolize, it displays extreme and unmanageable qualities if it is not united with its opposite. Without the mitigating influence of 1080, the power of 666 is that of the sun which, were it not for the protective atmosphere, would burn up the earth, or of the tyrant who rules for his own glory without consideration for the people, or of the rational principle where it entirely controls the mind and produces arrogance, self-delusion and madness. The Beast in Revelation signifies the total dominance of the number 666.

Being an essential part of nature, the number 666 demands recognition, and social or religious systems which attempt to ignore it are doomed to failure. A feature of the ancient Mysteries was the exhibition of sexual symbols, phallic pillars being erected at public religious festivals. The reason, given by Iamblichus in his book on the Mysteries, is that the ritual phalloi were 'signs of the prolific power which, through this, is called forth to the generative energy of the world. On which account also, many phalloi are consecrated in the Spring, because then the whole world receives from the Gods the power which is productive of all generation'.

This realistic attitude to nature was opposed by the early Roman Church, the phallic rites of the gnostics were put down and a moral code, based on the repression of human nature, was established, giving authority to the Church as its only interpreter. Christianity ever since has been more inclined to crusade against evil manifestations than to develop a philosophy for understanding their causes. Thus the Puritans at the Reformation, reacting against all that is represented by the number 666, attacked its symbols, such as crosses, monuments and religious images, the steeples of churches and the maypoles erected at celebrations of seasonal fertility, hoping thereby to suppress the corresponding elements in human nature.

The adoption of a moralistic or selective attitude to nature and religion leads to utopianism, the state of Utopia being a distorted image of the New

Jerusalem, in some ways its direct opposite. It is a product not of revelation but of human minds, reflecting the moral prejudices and fashionable theories of its time. As a creation of rationalism it is unable to comprehend all the opposite and apparently contradictory elements in nature, some of which must therefore be omitted. The effect on the excluded elements is like that which not being invited to the baptism had on the wicked fairy in the Sleeping Beauty story: she appeared in her most vindictive mood. Similarly, where due recognition is withheld from the number 1080, the spirit it represents turns stagnant and bitter, retires into the earth and becomes as a poisonous serpent or dark, malicious elemental. In the opposite situation, where the spirit symbolized by the number 666 is repressed, its male, authoritative attributes assert themselves in outbursts of violence and cruelty. On the other hand, if either of these symbolic numbers is excessively emphasized, the effect of the dominance of 1080 is anarchy and dissolution, traditionally known as death by water; and of 666 it is totalitarian rule and the worship of material products, leading to destruction by fire. To guard against those catastrophes is a function of the human sense of proportion, and to develop that sense is a function of the New Jerusalem. In the foundations of the divine city the numbers 1080 and 666 are evenly represented and united by their sum, 1746, the number of the Universal Spirit from which the negative and positive currents in nature both proceed.

The meaning of 666 is apparent from its gematria. It is that which comes from above, from God ($\pi\alpha\rho\alpha\ \theta\epsilon o\upsilon$, 666), and it is $\dot{\eta}\ \phi\rho\eta\nu$, 666, translated in Liddell and Scott's Lexicon as 'the heart, mind, understanding, reason' – the intellectual and rational part of the mind. Two other appropriate phrases, Divine Wrath ($\dot{o}\rho\gamma\eta\ \theta\epsilon o\upsilon$) and Weapons of God ($\ddot{o}\pi\lambda\alpha\ \theta\epsilon o\upsilon$) each number 665.

The elemental symbol of 666 is the fiery flying dragon, the opposite of the earth-bound serpent. It is the spirit of procreation described by the first hexagram of the I Ching oracle, on which Richard Wilhelm comments: 'The dragon is a symbol of the electrically charged, dynamic arousing fire that manifests itself in the thunderstorm. In winter this energy withdraws into the earth; in the early summer it becomes active again, appearing in the sky as thunder and lightning. As a result the creative forces on earth begin to stir again.' In his book on the oracle, *Change*, Wilhelm writes: 'In China dragons are not slain; rather their electrical power is kept in the realm, in which it can be made useful.' That is the traditional attitude to the dragon or beast, 666. Though its naked energy may be violently destructive it can not be described as evil, for it is the beloved mate of nature and the cause of life and beauty on earth.

The number of the beast in the Greek text of Revelation 13,18 is spelt in letters, $X\xi\varsigma'$ or 600, 60, 6. St John's words have stirred the imaginations of mystics and Apocalypse interpreters throughout our era. They clearly refer to some individual, a prince of this world, whose oppressive reign shall precede the appearance of the New Jerusalem. The problem of identifying the person with the number 666 has given rise to an extensive and curious body of literature, contributed to by many distinguished scholars. Irenaeus, the scourge of the gnostics, opened the debate in the second century by proposing several names or phrases with the number by gematria of 666, including $T\epsilon\iota\tau\alpha\nu$, Teitan, an archaic solar deity, and $\lambda\alpha\tau\epsilon\iota\nu o\varsigma$, Lateinos, a word of doubtful provenance meaning 'the Latin'. This name has been widely accepted, particularly among Protestant writers who have seen Lateinos as a fitting epithet of the Roman Pope. Cabalists among them have been ingenious in thinking up phrases with the number 666 which could be applied to the Church of Rome. Examples are $\dot{\eta}$ $\lambda\alpha\tau\iota\nu\eta$ $\beta\alpha\sigma\iota\lambda\epsilon\iota\alpha$, the Latin kingdom, and $\dot{\epsilon}\kappa\kappa\lambda\eta\sigma\iota\alpha$ $\dot{\iota}\tau\alpha\lambda\iota\kappa\alpha$, Italian church, and a claim attributed to the Pope, 'I am God on earth', $\theta\epsilon o\varsigma$ $\epsilon\dot{\iota}\mu\iota$ $\dot{\epsilon}\pi\iota$ $\gamma\alpha\iota\eta\varsigma$, also has the value 666.

Another link which has been perceived between Rome and 666 is that this number is the sum of the first six Roman numerals, I, V, X, L, C, $D = 1 + 5 + 10 + 50 + 100 + 500 = 666$. Nineteenth-century Protestant writers were pleased to observe that the Pope's triple crown bore the legend *Vicarius Filii Dei*, Vicar of the Son of God. Taking from that phrase the letters which serve as Roman numerals (*VICarIVsfILII DeI*), they pointed out that their sum amounted to 666.

Wealth, splendour and authority, attributes of the number 666, are undoubtedly possessed by the Roman Church, making it an easy target for anyone who finds significance in ascribing to it the number of the Beast. Prophetic assaults on those lines were being made against it long before Martin Luther's proclamation of the Pope as Antichrist and his Church as an institution of Satan. Luther's contemporary, John Foxe, author of the *Book of Martyrs*, was less certain about the Pope, believing that he was but a type of Antichrist and that the full title belonged to the nation of Turks. This comparatively moderate view of the Papacy was thoroughly out of fashion by the seventeenth century. Puritan theologians were so united in certainty that the Pope was the Beast, 666, that outrage was caused when one of their leaders, Richard Baxter, affirmed Foxe's view that the Pope was indeed a reflection of Antichrist but not exactly the real thing. His colleagues pointed out that Rome is unmistakably identified as Babylon in Revelation 17, where the Whore is seated on a beast with seven horns, interpreted later in the same chapter as seven mountains, which can be understood as the seven hills of Rome. Thus the seat of Antichrist was Rome, the seat of the Pope, and by equation the two were proved identical.

Typical of his time as a follower of this type of reasoning was Lord Napier, the mathematical genius, who invented logarithms in order to assist his calculations of the dimensions of the New Jerusalem. His conclusion was that Lateinos, the Pope of Rome, was the name of the man with the number 666. The theory remained popular up to the nineteenth century and was feebly reiterated in 1855 by the Rev. Reginald Rabbitt, who wrote a book on the number 666 abusing the Pope under the name Lateinos. Some years later Bishop Wordsworth in his commentary on the Greek New Testament continued the tradition of associating 666 with the Roman Pontiff by exhibiting a Vatican seal, bearing a device similar to the number of the Beast. The wittiest remark on the subject was Francis Bacon's, who said that if St Paul's description of the Antichrist in Thessalonians were issued to the police, the Pope would be arrested as the first suspect.

To all this the Pope's cabalists have made little answer, apart from showing that 666 is the number of Mahomet if spelt in Greek $M\alpha o\mu\epsilon\tau\iota s$, and their Protestant tormentors have now mostly lost interest in the matter. The rise of the ecumenical spirit and the waning appeal of biblical prophecies have caused a decline in the ranks of Apocalypse interpreters. A latter-day exponent, about sixty years ago, was the Rev. T. Simcox Lea, who interpreted 666 not as the Pope but as the Emperor of Rome, specifically Nero, because in Hebrew letters Neron Caesar has the value of 666, and the same name in Greek, $N\epsilon\rho\omega\nu$ $K\alpha\epsilon\sigma\alpha\rho$, is 1332 or 666 × 2. The great magician of his time, Aleister Crowley, observed that τo $\mu\epsilon\gamma a$ $\theta\eta\rho\iota o\nu$, the Great Beast, has the number by gematria of 666, and gained notoriety by adopting it as his magical name. Modern interpreters of the number of the beast have applied it to the name Hitler, which, if English letters are given numbers so that $A = 100$, $B = 101$, $C = 102$ etc., has the number 666; and if the name Stalin is written in Greek letters with the definite article as $\acute{o}$ $\Sigma\tau\alpha\lambda\epsilon\iota\nu$, it also bears the ominous number. In fact it is an anagram of $\lambda\alpha\tau\epsilon\iota\nu os$.

Behind this diverting nonsense is a real problem with a serious meaning: the problem of identifying the individual whom St John had in mind when he wrote, 'and his number is 666'. In solving it one must apply to St John's text the same cabalistic methods which, according to Irenaeus and other ecclesiastical writers, were practised by the gnostic masters and were therefore known to John himself. We are told to 'count the number of the beast'. The number 666 does not need to be counted or computed, for it is openly given, so evidently there is another number behind it which does need calculating. That number is in the final, key phrase of Revelation 13, 'And his number is 666', $\kappa\alpha\iota\ \dot{o}\ \dot{\alpha}\rho\iota\theta\mu os\ \alpha\dot{\upsilon}\tau o\upsilon\ \chi\xi s'$. The values of the words in that phrase are, $\kappa\alpha\iota = 31$, $\dot{o} = 70$, $\dot{\alpha}\rho\iota\theta\mu os = 430$, $\alpha\dot{\upsilon}\tau o\upsilon = 1171$, $\chi\xi s' = 666$. The sum of these numbers, and therefore the value of the whole phrase, is 2368, which is the number of $'I\eta\sigma o\upsilon s\ X\rho\iota\sigma\tau os$, Jesus Christ. The esoteric meaning of the whole verse is therefore: 'Here is wisdom. Let him that hath understanding count the number of the beast: for it is the number of a man: 2368, Jesus Christ'.

It is only at first sight that this interpretation seems shocking and paradoxical. When it is considered in context, as illustrating the gnostics' dispute with the institutionalized Church, its meaning is readily apparent. The main difference between the two sides was over the nature of Christ's divinity. The gnostics affirmed that the spirit of Christ was divine, but they thought it absurd to worship the body of Jesus or any material image. They criticized the Church for its emphasis on the historical figure of Jesus, for proclaiming the divinity of his human body and for making him, in modern terms, the object of a personality cult. Their own view of the matter was that Christ was a redeeming spirit, a renewed archetype, whose coming recharged the atmosphere and opened a new area of human understanding. They believed in personal communion with that spirit and aspired to mystical union with it, thus coming into conflict with the Church and the monopoly it claimed on religious instruction. Most blasphemous, to the gnostics' way of thinking, was the erection of an idol, the image of the wounded man on the cross, as an object of compulsory Christian worship. St John therefore stigmatized that idol as the Beast 666, the number of unbridled solar power, using the same words to impart to those with understanding the number of its name, 2368.

St John's account of the corruption of Christianity is hidden behind the allegory of the beasts in Revelation 13. The prophet sees the first beast rising out of the sea. This is the great event which gave birth to Christianity: the dawn of a new sun and the rising up of a new god or archetype from the subconscious mind, symbolized by the sea which is also *mare* or Mary.

The second beast then appears, coming out of the earth. 'He had two horns like a lamb, and he spake as a dragon.' He commands everyone on earth to worship the first beast 'which had the wound by the sword, and did live'. In the name of the first beast he makes war on the saints and overcomes them and establishes his authority throughout the world. He deceives people by means of the wonders he can perform through the power of the first beast, persuading them to make an image of the wounded beast and to worship it. Those who will not do so are killed, and anyone who is not stamped with the mark, name or number of the beast is excluded from society and may not buy or sell in the market.

St John's meaning is plainly in accordance with the course of events as perceived by the gnostics. The second beast comes from the earth and is therefore a material creature, the body of the Church. It manipulates the power of the first beast, the spirit of the Redeemer, promotes its influence and identifies itself as its agent. Having gained power, it consolidates its hold by suppressing the gnostic saints and by claiming credit for all miracles performed in the name of Christ. Finally it sets up an idol, the wounded figure of Jesus Christ, cajoles people into worshipping it and excommunicates those who will not do so. In the last words of the chapter is spelt out the esoteric number of the first beast, 2368, the number of the man on the cross.

St Paul, whom the gnostics claimed as one of their number, made a similar comment in the first chapter of his Epistle to the Romans, condemning those who 'changed the glory of the uncorruptible God into an image made like to corruptible man . . . who changed the truth of God into a lie, and worshipped and served the creature more than the Creator'.

Throughout the Book of Revelation scenes of corruption and horror alternate with images of paradise. A remarkable feature is that, although the principal figures in the different scenes are apparently the exact opposites of each other, they are also in many ways comparable, as if representing different sides of the same coin or different aspects of a number. Scholars have commented on the similarity between, for instance, the word used for the Whore, ἡ πορνη, and the Bride, ἡ νυμφη, and between the Beast, το θηριον, and the Lamb, το ἀρνιον. These characters belong to two different cities, Babylon the great, luxurious and decadent, moving towards its doom, and new-born Jerusalem with its fresh springs and walls of sparkling crystal. In their outward forms these two cities are as different as could be, but we learn through gematria that in fact they are really one, for

$1285 = Ba\beta \upsilon \lambda \omega \nu$, Babylon
$= \dot{\eta} \; \dot{a}\gamma \iota a \; \pi o\lambda \iota s, \; ' I\epsilon \rho o\upsilon \sigma a\lambda \eta \mu$, the holy city Jerusalem.

The prophet understands that human nature is always and everywhere the same, and that the simple, innocent people of Jerusalem are no different in nature from the corrupt citizens of wealthy Babylon; they are merely born during different stages of civilization and behave accordingly. Excavators of ancient cities constantly discover that the same site has been occupied over several cycles of civilization. In the beginning it is like Jerusalem, a settlement around a sacred spring or shrine of the goddess. As it flourishes so its affairs grow more complicated. Law courts, state temples, prisons, ports, palaces and the other civilized institutions come into being, and finally the city is no longer in scale with its surroundings but has become a bloated parasite, living off the produce of other lands. At that stage it is Babylon or Plato's Atlantis, a rich, indolent city ripe for destruction. Throughout the whole career of the city its site is the same, and so are the natures of the people born into it. The apparently radical difference between Jerusalem and Babylon is therefore but an illusion of time, and in the philosopher's language the same number, 1285, applies to them both. It is also the number of the deity whose shrine was the cause of the original settlement. The name given her by the Orpheans (according to Proclus) was the Life-bearing Goddess, $Z\omega o\gamma o\nu o s \; \Theta \epsilon a$, 1285.

In the gnostic writings included in the New Testament can be found many cases where two names, signifying morally opposite principles, have the same number and can therefore be seen as two aspects of the same type. St Paul in II Thessalonians 2, 3, warns of a coming antichrist, the Man of Sin, $\dot{o} \; \dot{a}\nu \theta \rho \omega \pi o s \; \tau \eta s \; \dot{a}\nu o\mu \iota a s$, 2260. This is also the number of the idol set up by the second beast in Revelation 13, called the Image of the Beast, $\dot{\eta} \; \epsilon \dot{\iota} \kappa \omega \nu \; \tau o\upsilon \; \theta \eta \rho \iota o\upsilon$. The coincidence of numbers here is obviously congruous; but furthermore 2260 is the number of $\dot{o} \; \upsilon \dot{\iota} o s \; \dot{a}\nu \theta \rho \omega \pi o\upsilon$, the Son of Man, who comes on a white cloud in Revelation 14, armed with a golden sickle to reap the harvest of the earth. The Son of Man being a epithet of Christ, here again is the paradox of both Christ and Antichrist bearing the same number. Here also, as St John said, is wisdom. From the gnostic sages we learn that energy is morally neutral and that the form of it which entered the world two thousand years ago had two different effects. As the spirit of Christ it stimulated minds and brought a renewal of prophecy and culture; but it also gave life to an Antichrist, intolerant and power-engrossing, who persecuted the saints from his seat in the imperial city. We also learn that the notion of good and evil is irrelevant to nature, which works out its cycles with no regard for human fads and preferences; nor does it apply to the eternal types and tendencies which are described by symbolic numbers. These can not therefore be judged good or evil in themselves, but are considered philosophically through their relationships, whether they are moderate or excessive, balanced or disproportionate, aesthetically satisfying or unpleasant to the senses. For deciding these matters the human mind is equipped with a sense of order and proportion, and this inherent sense can best be developed, according to the traditional recipe, by means of our present study, the structure and symbolism of number.

1746, the number of fusion

The number 1746 is a symbol of fusion between the negative and positive forces in nature, 1080 and 666, of which it is the sum. It is the number by gematria of το πνευμα κοσμου, the Universal Spirit, which is the combination of those two forces, and it is the number of the name given to the light which streamed out of the Holy of Holies at seasons when the electric currents of the atmosphere were fused together with the magnetic energies of the Earth Spirit at the Temple of Jerusalem. The name of that light was the Glory of the God of Israel, ἡ δοξα του θεου Ἰσραηλ, 1746.

1746 is the number of that fertilized seed from which, according to ancient philosophy, the whole universe grew up like a tree. Modern cosmologers are inclined to see things similarly, though not in such organic terms, choosing characteristically a more violent image for the process, a 'big bang'. The primal germ is referred to in the three Synoptic Gospels as the 'grain of mustard seed', κοκκος σιναπεως, 1746. It is called the least of all seeds, and from it sprang the universal tree and the living things that inhabit it. 'And the birds of the air sheltered in its branches' – Luke 13, 19.

The allegory of the grain of mustard seed is of particular interest in St Mark's version, chapter 4, because there it is followed by a verse referring to an esoteric doctrine behind the outward sayings of Jesus: 'But without a parable spake he not unto them: and when they were alone, he expounded all things to his disciples.' The grain of mustard seed is shown by gematria to have been an early Christian symbol of their Founder's secret teachings, for its number, 1746, is also that of το κεκκρυμενον πνευμα, the Hidden

Spirit, ὁ θησαυρος 'Ιησου, the Treasure of Jesus, and several other phrases of like meaning. The purpose of those teachings, to prepare the way for divine rule on earth, is represented by the same number, for

1746 = Jerusalem, the City of God, 'Ιερουσαλημ, ἡ πολις θεου
     = the City of the Saints of God, ἡ πολις ἁγιων θεου

In the course of prolonged studies in gnostic and New Testament gematria the following terms and phrases have been discovered, illustrating the symbolic character of the number 1746.

1746 ≐ the Universal Spirit, το πνευμα κοσμου
     = Grain of Mustard Seed, κοκκος σιναπεως
     = the Glory of the God of Israel, ἡ δοξα του θεου 'Ισραηλ
     = the Hidden Spirit, το κεκκρυμενον πνευμα
     = the Treasure of Jesus, ὁ θησαυρος 'Ιησου
     = the Divinity of Spirit, ἡ θεοτης πνευματος
     = the Spiritual Law, ὁ νομος ὁ πνευματικος
     = the Chalice of Jesus, το ποτηριον 'Ιησου

This number appears to have a special association with the age of Pisces and the Christian era. It pertains to epithets of Christ as the universal seed germinating in the matrix of nature:

1746 = Son of Virgin Mary, υἱος παρθενου Μαριας
     Offspring of a Virgin's womb, γεννημα γαστρος παρθενου
     Jesus, Child of Holy Mary, 'Ιησους, παις ἁγιας Μαριας
     Jesus, Infant Lamb of Mary, 'Ιησους, ἀρνιον παιδιον Μαριας
     Precious Pearl of Mary, τιμιος μαργαριτης Μαριας
     Emmanuel the Son of Mary, 'Εμμανουηλ ὁ υἱος Μαριας

Adjacent numbers have gematria of similar meaning. 1747 is the number of the Holy Spirit and the Bride, who are united at the end of Revelation (22, 17), and it also gives phrases to describe Jesus Christ and his mystery teaching.

1747 = Holy Spirit and the Bride, ἁγιον πνευμα και ἡ νυμφη
     the Spirit of God on Earth, το πνευμα θεου ἐπι γαιης
     Christ the Kingdom, Χριστος ὁ Βασιλεια
     the Divinity of the Kingdom of Jesus, ἡ θεοτης βασιλειας
     'Ιησου

Fruit of the Vineyard (Mark 12, 2; Luke 20, 10), καρπος ἀμπελωνος

Knowledge of God, γνωσις θεου

Seven Spirits of God (Revelation 5, 6), ἑπτα πνευματα θεου

Mysteries of Jesus, μυστηρια Ἰησου

the True God of Life, ὁ θεος ἀληθινος ζωης

1745 gives the Trinity, Father, Son and Spirit, the twelve apostles and some traditional names of the Virgin goddess.

1745 = Father, Son, Spirit, πατηρ, υἱος, πνευμα

     the Twelve Apostles, οἱ δωδεκα ἀποστολοι

     Queen of Heaven, Βασιλισσα οὐρανον

     the Mother of Grace, ἡ μητηρ χαριτος

     the Mother of All Things, ἡ μητγρ παντων

     the Virgin, Pearl of the Kingdom, ἡ παρθενος, μαργαριτης βασιλειας

Images of 1746 are the Word that was in the beginning and the fertilized ovum. In the alchemist's retort, as in nature's womb, the opposite elements are fused together, and the fruit of their union is the Universal Spirit, 1746.

The important part which this number plays in the New Jerusalem scheme is seen in figure 12, where 1746 is the height of the Vesica containing the canonical circle of radius 5040.

Sacred names from the dimensions of the New Jerusalem

The existence of so many Christian sacred names and phrases reflecting by gematria the principal numbers of the New Jerusalem diagram suggests that the plan of the Holy City was the prototype from which the forms of Christianity were largely derived.

One of the many symbolic functions of the diagram was astrological. The start of the Christian era coincided with the dawning of the age of Pisces, wherefore Jesus was known esoterically as the Fish. Corresponding to Pisces among the Greek gods was Poseidon, god of the sea. Thus in Revelation 13 the first beast (identified in the section on 666, above) arises from the sea to become the astrological dominant of the new age. During his reign he is Father of the World, πατηρ κοσμου, 1289, and the same number emerges from the gematria of the names and symbols of the Piscean ruler.

$1289 = \tau o\ \acute{o}\nu o\mu a\ \text{'}I\eta\sigma ou$, the Name of Jesus

$= \acute{o}\ \grave{\iota}\chi\theta us$, the Fish

$= \acute{o}\ \Pi o\sigma\epsilon\iota\delta\omega\nu$, Poseidon

In Revelation 22 is displayed the Tree of Life 'which bare twelve manner of fruits, and yielded her fruit every month.' The cabalistic Tree of Life is another image of the New Jerusalem, its twelve lunar fruit corresponding to

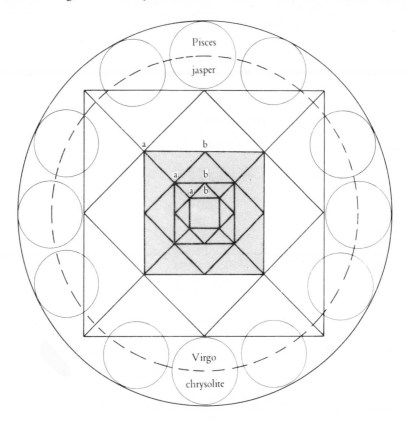

Figure 70. The square and circle of the New Jerusalem diagram form together an astrological chart for the age of Pisces and a table of magical correspondences, setting out in order the names and attributes of the twelve universal powers along with the appropriate tribes of Israel, angels, apostles etc. The correct order and positions of these names being uncertain and a matter of dispute among scholars, no attempt is here made to supply them. The use of such a chart is to give necessary information to practitioners of ritual magic.

Also demonstrated in the diagram is a geometrically satisfying means of dividing a square area equally between tribes (*a, a, a* . . .; *b, b, b* . . .) up to a central pivot.

the twelve moon circles in the diagram, each of diameter 2160. Here they represent the twelve astrological 'months' of the Great Year in which the sun completes its course through the zodiac. One such month lasts 2160 years.

The twelve-fold symbolism of the New Jerusalem consists of:

12 gates attended by 12 angels and inscribed with the names of the 12 tribes of Israel;

12 foundations, ornamented with 12 kinds of precious stone and containing the names of the 12 apostles.

Some of these elements denote particular geometric shapes, angels perhaps being angles, and the names are added to the diagram to make it efficacious for divination, as a talismanic figure or astrological chart. Behind the various orders of supernatural creatures in the chapters of Revelation is a table of magical correspondences which students of esoteric science have attempted to reconstruct. The question of how the twelve tribes of Israel, listed by St John in Revelation 7, should be correlated with the signs of the zodiac, and how they should be placed with the precious stones and other elements on the New Jerusalem chart is discussed by Dr R.H. Charles in his *Commentary on the Revelation of St John*; but his astrological attributions differ from those given in Jewish writings, so there is no certainty in the matter. Less controversial is the order of the New Jerusalem's gem stones in relation to the twelve signs. These are given by Athanasius Kircher in *Oedipus Aegyptiacus*, to whose list below are added the traditionally corresponding Greek gods.

Aries	amethyst	Athena
Taurus	hyacinth	Aphrodite
Gemini	chrysoprase	Apollo
Cancer	topaz	Hermes
Leo	beryl	Zeus
Virgo	chrysolite	Demeter
Libra	sardius	Hephaistos
Scorpio	sardonyx	Aries
Sagittarius	smaragdus	Artemis
Capricorn	chalcedon	Hestia
Aquarius	sapphire	Hera
Pisces	jasper	Poseidon

In the course of the Great Year the sun goes backwards through the zodiac, moving from Pisces into Aquarius. That is followed by the order in which the precious stones are named in Revelation and in which they are here placed on the diagram. In this form the New Jerusalem becomes the nativity of Christ and the Christian era.

At the bottom of the chart, opposite Pisces, is Virgo, by which arrangement is figured the birth of the Fish god from the Virgin. The sum of Pisces and Virgo, 'Ιχθνes, 1224, and ἡ Παρθενos, 523, is 1747, equivalent to that 'number of fusion' which occurs most frequently in Christian gematria.

Adding together the values of the precious stones at the top and bottom of the chart reveals the number of an epithet of Jesus, Son of Man.

501 = ἰασπις, jasper
1689 = χρυσολι θos, chrysolite
2190 = υἱos ἀνθρωπου, Son of Man.

The nature of the New Jerusalem and the means by which it becomes manifest are indicated by its gematria. Its Greek name, ἡ καινη 'Ιερουσαλημ, has the value of 961, equating it with ὁ οὐρανos, 961, heaven. It is indeed that pattern in the heavens which Plato wanted to bring down to earth and establish as a universal standard of proportion. The secret discourses, τα ἀπορρητα, 961, which were a traditional feature of the initiation process, may have been, in accordance with the number, on the subject of the heavenly city. That number also indicates the manner of the New Jerusalem's appearance, suddenly as a revelation from above; for 961 is the number of το διos βελos, the divine thunderbolt.

6 The ancient and future cosmology

THE OVERALL PURPOSE of modern speculative physics is to construct a Grand Unified Theory demonstrating mathematically the generation of the physical universe and its underlying dynamics. Ancient scientific meta⁄physics had a similar but wider function. It attempted to describe not merely the nature of the physical universe but human nature also, and in the same terms, linking the two together as macrocosmic and microcosmic aspects of the one primordial act of creation. The universe, human nature and the mind of the Creator were made commensurable by number, which Plato called the 'bond' holding all things together. Of number was constructed the sacred Canon, the first, most abstract symbol of reality, formalized in the New Jerusalem diagram. That diagram is here identified as the ancient world⁄image or cosmology.

This chapter is about cosmology, its nature and influences, the ancient perception of its power to shape societies, its present function and the options which are available in adopting a suitable cosmology for these and future times.

A cosmology in the present sense is an overall view of the world as formed by individuals and by societies as a whole. Everyone has a personal cosmology, compiled from their experiences, prejudices, conclusions, beliefs and wishes, certain features of which are held more or less in common by people of the same culture. The common features in the cosmologies of individuals constitute, and also reflect, the accepted world⁄view of their societies.

The fundamental importance of cosmology lies in the fact that the images which a person or society projects upon the universe condition their whole experience of life. The universe by definition is all⁄inclusive and it can also be termed reflexive, implying in it a tendency to respond positively to any conceivable idea applied to it. Thus any system of belief tends to attract evidence and phenomena which seem to confirm it. A cosmology for that reason is a powerful artifact. It serves to actualize the reality which it purports to describe.

Because it shapes the reality he is accustomed to, a person's cosmology becomes his most precious and jealously guarded possession. It is a commonplace of psychoanalysis, confirmed by everyday experience, that people tend to ignore or angrily reject unfamiliar ideas which have no place in their mental cosmologies. Similarly with societies, unorthodox views and observations which run counter to the established cosmology are not gladly received and may well be suppressed.

The ideal cosmology is the picture of the world which most closely resembles its original. Yet between the original and any rational representation of it there is a dimensional gap. The universe as an organism is a creature of paradox, never entirely predictable, whereas a rational cosmology must be self-consistent. Plato in *Parmenides* showed that for every general statement which can be made about the universe the opposite statement is equally tenable, thus invalidating all approaches to cosmology which fail to allow the upholding of two contrary ideas at the same time. That necessity is symbolized by the interlaced square and circle at the foundation of the New Jerusalem.

For a cosmology to be successful and lasting it needs to be inclusive, capable of reflecting every possible type of human experience, physical, mental and those which lie in between. The picture of reality which emerges through neutral observation of its various manifestations is different in quality from that obtained through any system of belief, whether it be called science, religion or philosophy. As noted by that shrewd cosmologist, Charles Fort, all such systems are based on 'exclusionism', meaning that they necessarily disregard phenomena which challenge their basic premises. The cult followers of Darwin, Freud, Marx, Einstein and all other systematizers are compelled for the sake of consistency to ignore or devalue testimonies which do not suit their book; and the Church (though avowedly based on faith and therefore less vulnerable to incidents of being proved wrong) has nevertheless felt threatened by every new discovery which went against its temporary dogma.

The result of exclusionism is an incomplete cosmology, restricting both experience and understanding and thus incapable of long endurance. However staunchly it may be defended by its adherents, reality is constantly at work to undermine it. The aspects of nature and humanity which it neglects are those which cause trouble. Like portents unheeded, they become more and more insistent in drawing attention to themselves until the excluding system gives way or is adapted to accommodate them. An illustration of the type of reaction produced by a one-sided cosmology is given by Oscar Wilde in *De Profundis*, written from prison in the years following his disgrace.

> I remember when I was at Oxford saying to one of my friends as we were strolling round Magdalen's narrow bird-haunted walks one morning in the year before I took my degree, that I wanted to eat of the fruit of all the trees in the garden of the world, and that I was going out into the world with that passion in my soul. And so, indeed, I went out, and so I lived. My only mistake was that I confined myself so exclusively to the trees of what seemed to me the sun-lit side of the garden, and shunned the other side for its shadow and its gloom. Failure, disgrace, poverty, sorrow, despair, suffering, tears even, the broken word that comes from lips in pain, remorse that makes one walk on thorns, conscience that condemns, self-abasement that punishes, the misery that puts ashes on its head, the anguish that chooses sackcloth for its raiment and into its own drink puts gall: — all these were things of which I was afraid. And as I had determined to know nothing of them, I was forced to taste each of them in turn, to feed on them, to have for a season, indeed, no other food at all.

Wilde's experience typifies the consequences of a personal cosmology which rejects any of life's aspects. The same lesson applies to public cosmologies, those which belong to whole nations or cultures. To the extent that they fail to recognize and allow for all possible elements in human experience they are inadequate, producing friction between convention and reality.

The rulers of our western world today uphold no formal cosmology and tolerate free speculation about the nature of things. Yet at the root of all our institutions there is an undeclared cosmology, a conglomeration of received theories and assumptions which constitute the dominant orthodoxy. In order to perceive its character and inherent tendencies it is necessary to be detached and view it from outside itself; but to those of us who are part of it, educated within its framework, that is no easy task. An external viewpoint is required, and that is provided ready-made by the traditional form of cosmology which takes human nature as a constant reflection of divinity and values its interests accordingly.

The view of the present which unfolds is of humanity expelled from the centre of the former organic, divinely created cosmos, and placed, irrelevant and purposeless, among the lifeless fragments of a universe constantly expanding. That image may as easily be credited, or as freely rejected, as any other. Inseparable from it, however, and more easily evaluated, are its effects. As the prevailing image of modern scientific cosmology, it gives life to subsidiary images and their corresponding phenomena – the expansionist state, the engrossing corporation, technological developments inimical to human interests – which distinguish the present age from those preceding it. Among the side-effects of modern cosmology is an exaggerated respect for inventiveness, along with aggressive ambition. These and other such expressions of the number 666 operate at the expense of that other side of human nature, its earthly element, which eschews innovations and finds comfort in a traditional order with established customs and continuity of culture.

The lack of a declared humanistic basis to modern cosmology has deprived the present age of the proper yardstick for discerning aberrant developments. Thus policies and institutions are free to flourish which, from the traditional standpoint, appear vain, inhuman and oppressive. Generated by them are equally extreme reactions in such forms as anarchism, nationalism, sectarianism, fundamentalist religion and symp-toms of alienation ranging from apathy to terrorism. In traditional terms it would be said that the powers of 666 and 1080 have turned their backs on each other and are competitively engaged in manifesting their grossest forms. The balance thus created is that unstable variety which consists of lurching between extremes.

A mundane comparison is with a vehicle swerving out of control. In that situation the passengers call out for a firm hand at the wheel; and similarly, when societies are in crisis and destruction threatens, the demand arises for a strong leader or dictator. Thus, as Plato shows in the *Republic*, a society without a basic standard, with no means of discriminating between harmonious and disproportionate forms, is destined to fall under tyranny.

The alternative, more lasting remedy is supplied by traditional philosophy: to discern the nature of the two forces which confront each other across the split down the modern world-order, and to effect their reconciliation through the medium of that constant standard, the ancient canon of proportion. One of its symbols is the *axis mundi* or world-pole, the central feature of traditional cosmology, which stands fixed and firm amid the ever-changing universe. Its function is like that of a magnetic rod which gives pattern and order to the particles within its field of influence.

Consideration of the cosmological types available for the future reduces their number to three principal categories. First is the modern western variety, based on no fixed standard in nature, formally undefined and shaped by conflicting opinions and interests. Its benign product is the 'open society' lauded by Sir Karl Popper and fellow liberals, which has many attractions but one fatal flaw: an inherent instability which sooner or later overturns it. The second type represents a decline of the first, to the point where standards are found necessary to maintain order, and these are provided by some artificial structure such as personal tyranny, dogmatic religion, 'scientific atheism', Fascist economics or Marxian historical theory. Such impositions are both irksome in practice and inadequate frames to the world and human nature. The security they promise is never actually experienced, for as mere human concepts their eventual destruction or chaotic collapse is assured.

The third possible type of cosmology is as described in this book, the traditional model symbolized by the New Jerusalem. Its character and outline have already been examined. As a synthesis of the proportions and harmonies in the field of number, it depicts the essential structure of the universe and the human mind alike, uniting them both within the comprehension of reason. The ancient philosophers regarded it as the closest approximation to the nature of things that can rationally be conceived. Its potential for wide acceptance and long endurance are not in doubt. The important question is as to the effects it has on people and societies under its influence.

In the New Jerusalem diagram the pole of the universe passes through the centre of the earth which is therefore at the heart of things. That is an optional pattern in terms of scientific astronomy, for in the world of relativity the conceptual centre of the universe is wherever one chooses to locate it. As a cosmological proposition its corollary is that earth-born humanity is central to the universal scheme, and that human nature is no chance by-product but an essential factor of Creation. Since actions at the capital or hub affect the provinces and periphery, the implication is that human deeds and thoughts have influence on the world at large. This view appeals to moralists as tending to promote a sense of purpose and responsibility in individuals, and it is also attractive to mystics for giving sanction to many aspects of experienced reality which the present world-view excludes. It promotes the concept of a reflexive universe, referred to above, which is subjectively but widely evidenced by incidents of telepathy, precognition, coincidences perceived as significant, 'mind over matter' and other such items of common experience. Thus the first effect of traditional cosmology is to restore and sanctify humanism. As illustrated by the medieval and renaissance cosmographies in which the world-centre is the navel of the archetypal man stretched across the whole universe, human nature is understood as the microcosm, partaking in the divine nature and so worthy of being held as a constant standard of reference.

Acceptance of human nature as a true standard (a concept bravely but briefly revived by Pelagius in the fifth century) does not of course imply that any one of its manifest examples is perfect or infallible. Just as the forms of nature – the rose, the crystal and so on – reflect an ideal symmetry which no individual among them ever achieves, so it is with humanity. In deriving all the forms of nature from ideal, unmanifest prototypes, traditional philosophy reverses the evolutionists' notion of human ascent from lower creatures and inculcates the opposite myth, that we are descendants from a divine creation and may properly aspire to re-enter the primeval paradise. That point of view is incompatible with the outlook of modern science; nor does it commend itself to the authorities of church and state who flourish by virtue of the belief that human nature, being flawed, needs constant suppression. (Such a belief is in the interests of tyrants, allowing them to blame the populace for disturbances which might otherwise be attributed to their own policies.) Traditional cosmology therefore calls for a different form of science and a different approach to government from those which obtain today.

The style of government which occurs naturally under the influence of the canonical world-view is a hierarchy representing the order of the heavens. The sacred ruler, king or high priest has a largely ceremonial function. In the social order he corresponds to the sun in the celestial hierarchy and imitates in his daily ritual its journey through the skies. At his coronation he is charged with the mystic energy, known as *prana* and by many other names, which he transmits through the realm, making it prosperous and fertile. An alchemical symbol of his function is a golden crown lodged in the midst of a tree where the branches meet the trunk – as enacted by King Charles II when he took refuge in an oak. From his position mid-way up the symbolic tree the king draws down the power of 666 through its leaves, and is also sustained by the energy of his earth-bound subjects, the element numbered 1080 in the cosmic equation, rising upwards through its roots. In him are fused together the two opposite elements, and thus is generated the sacred energy which like a benediction pervades the entire country.

In the old state cosmologies the capital city was placed around the navel of the earth, with the citadel and temple at the centre and the king enthroned at the heart of all. This pattern served also as a mental cosmology for individuals; for as the king is to the state so is each person in relation to his surroundings, placed at the centre of his universe and responsible for its ordering. The corresponding political pattern is of a central unifying authority, useful for defence and for balancing famines and gluts in the provinces. Its powers and functions are ritualized, regulated by the canon of proportion applied to statecraft. The intention is to allow the greatest possible degree of autonomy to provinces, communities and individuals. As an illustration of how well this was once achieved in imperial China, it was said that people in remote villages were so little disturbed by the emperor's officials that not even news of his death reached them, and they were still worshipping the same emperor two or three reigns later.

The ancient art of harmonious government was based on the perception of how the conduct of rulers sets the pattern for the whole state. Thus the actions of the king and court were circumscribed and ritually attuned to the cosmic order. The caste of Guardians whom Plato appointed to govern his Republic were to be subject to similar disciplines. Natural disasters and popular disaffection were seen as having the same type of cause, some error in the style of government which upset the balance of forces and produced friction between earth and heaven. Instead of imposing their own invented system and blaming their subjects if they failed to appreciate it, ancient rulers and their advisers were bound to observe and respond to every natural portent, including the changing temper of the people.

In the traditional societies of old China and the East all types of unusual happenings were noted by the local authorities and reported to central government. The noted incidents might include sightings of aerial dragons, strange lights, phantoms and unexplained creatures, odd animal behaviour, monstrous births, meteorological freaks and other such wonders and prodigies. Apparently meaningless in themselves, such things, if widely reported in the same period, were taken as symptoms of psychic unrest presaging some such social or natural upheaval as riots or earthquakes. The meaning of the symptoms was decided upon by astrologers and the appropriate adjustments were made to the style of government. Hellmut Wilhelm in *Change*, his commentary on the *I Ching*, gives examples of flaws in government which, according to the Chinese, have certain consequences in nature. Offences against ritual, the appointment of unworthy persons, the dismissal of the worthy and listening to slander cause outbreaks of fire and strokes of lightning. Excess and wastefulness in the administration bring about heart and abdominal disease, dust storms and earthquakes. Another ill, the black evil, associated with careless observation of marriage rites and the wrong relationship between emperor and people, produces nervousness and diseases of the ear, long cold spells and deaths among animals. In the ancient collection of Chinese texts, brought together in 1050 BC and known as the *Great Law*, the importance of observing omens is clearly explained:

It is the duty of the government all the time to watch carefully the phenomena of nature, which reflect in the world of nature the order and disorder in the world of government. The government is bound to watch the phenomena of nature in order to be able at once to change what is in need of change. When the course of nature runs properly, it is a sign that the government is good, but when there is some disturbance in nature, it is a sign that there is something wrong in the government. With the help of fixed tables it is possible to learn from the disturbance in nature what is the sin that caused it. Any disturbance in the sun accuses the emperor. A disturbance around the sun accuses the court and the ministers. A disturbance in the moon accuses the queen and the harem. Good weather that lasts too long shows that the emperor is too inactive. Days which continue to be cloudy show that the emperor lacks understanding. Too much rainfall shows that he is unjust. Lack of rain shows that he is careless. Excessive cold shows that he is inconsiderate of others, stormy winds that he is lazy. A good harvest proves that all is well, a bad harvest that the government is at fault.

This old way of looking at things and the old order of values were so different from those orthodox today that at first one can hardly believe that they were once commonly accepted by members of the same human species as ourselves. Yet no one supposes that human nature has changed in recent ages, and the circumstances of life on earth are much the same as they have always been. The days when civilizations were governed by the cosmic canon of proportion are not therefore hopelessly beyond recall; they are separated from us merely by conventions of thought. If we find it in our interest to seek once more a standard of affairs in human and universal nature, there is nothing to inhibit us from doing so. It would be nothing more than a return to orthodoxy.

No one with any sense, who appreciates a quiet life, wants to stir up a hornet's nest by challenging the powerful institutions of the present, or even to waste time campaigning and propagandizing for an alternative cosmology. Radical changes involving the replacement of one world-view by another come in their own good time, when circumstances and human minds are ready for them. The process seems similar to an invocation. The New Jerusalem is not a constructed world-image but a pre-existent archetype which reveals itself, as to St John on Patmos, where there are minds prepared to receive it. As an individual, one can best prepare for changes in the future by taking the New Jerusalem image as a model for structuring the mind, thus acquiring an inclusive personal cosmology which gives insight into the reality behind the apparent forms of the present. From this perspective one can see historical processes as parallels to those of alchemy, the modern age being evidently one of *separatio*, often represented in alchemical texts by the dismemberment of a body, in which the elements lose balance and cohesion and split apart from each other. In such terms may be described the rift in the modern world-image, the separation of 666 and 1080 and the clash of ideologies representing the opposite principles.

The next stage in the alchemical model of history, following *separatio* and the clear discernment of opposites, is *coniunctio*, where the dispersed, purified elements come together to form a structured organism. From this union there is an offspring who, as in the Christian myth, dies and is resurrected. This completes the first stage of the alchemical process which, historically interpreted, leads to restoration of the age of gold.

A similar analysis of modern times, tending to an optimistic conclusion, occurs in the *Alchemical Studies* of C.G. Jung.

> The balance of the primordial world is upset. What I have said is not intended as a criticism, for I am deeply convinced not only of the relentless logic but of the expediency of this development. The emphatic differentiation of opposites is synonymous with sharper discrimination, and that is the *sine qua non* for any broadening or heightening of consciousness.

The dichotomy or loss of balance in the modern world-order is undoubtedly perilous and may well have a cataclysmic outcome. Jung's belief in salvation through the 'broadening or heightening of consciousness' is echoed by many spiritual leaders today, but it is a somewhat general formula, leaving room for further inquiry into the possible shape of affairs following such a change. We have previously contrasted the different types of cosmology: those which are based on a universal standard, rooted in nature, and those which adhere either to no declared standard or to one that is artificial and temporary. The second category is at present in the ascendancy, so any change in consciousness involving a radical shift in cosmology may well restore the first category to favour. In that case, the traditional canon of proportion and its cosmological image, the plan of the New Jerusalem, are as capable today as ever they were in the past of providing the required standard.

Heavenly Jerusalem, its invocation and modern function

Idealism is currently out of fashion; the tenor of modern philosophy is against it, and against the entire way of thought illustrated in this book. Common criticisms of idealism are either that, though harmless, it is unreal and impractical, or that it is a cause of dangerous fanaticism. The second objection has a good deal of force; the idealism of unstable characters can turn to monomania with unpleasant consequences. But it is not idealism proper which is thereby discredited, merely its perversions. And the first assertion, that idealism is impractical, is no more self-evident than the opposite opinion, here affirmed, that it is in fact the only practical means of restoring balance to the world.

In the previous chapter were pointed out the advantages of an objective cosmological standard in human affairs, and it was argued that the identification and adoption of such a standard is the one realistic option for the future. That necessary standard has traditionally been provided by the geometers' diagram of an ideal city, Heavenly Jerusalem, and, behind the diagram, by an esoteric code of number, the Canon, representing the combined formulas of natural growth and motion. The outcome of these studies in the Canon, the ideal City, the musical harmonies of Creation and other expressions of the universal archetype is the conviction that the type of understanding which develops through them is destined to expel the gross, destructive thought-forms of modern materialistic philosophy. That issue is not only desirable but, since human mentalities adapt naturally to the requirements of their time, to be expected. It implies revival of the traditional form of science, most appropriate to these present times, which has for its object the invocation of the Heavenly Jerusalem.

The nature of the Heavenly Jerusalem, its form, dimensions and symbolism, have already been discussed at length, but little has been said about its material reflection, the actual city of Jerusalem in the Middle East. The Heavenly City, as the pattern of ideal order, has of course no exclusive association with any particular spot on earth. But it so happens that Jerusalem is now most widely acknowledged as the *omphalos* or sacred centre of this planet. To followers of the three most powerful religions of the West, Christian, Jewish and Muslim, it is a shrine of unique importance. It was the scene of Abraham's sacrifice, the Crucifixion of Jesus and Mohammed's ascent to heaven. The waters of the Flood arose from beneath its central rock and afterwards subsided there. Its legends, the features of its sacred geography and its very name (meaning in Hebrew Peace and Wholeness) identify it as the earthly type of the Heavenly City. This identification is strongly supported by history. Throughout our era Jerusalem has been the goal and inspiration of innumerable chiliastic and idealistic movements, from the Christian crusades to modern Jewish Zionism. The declared aim of esoteric groups such as the Freemasons and the Knights of St John is the rebuilding of Jerusalem's Temple, by which is symbolized restoration of the traditional code of philosophy and the ideal reordering of human society.

The present situation in Jerusalem is that the Jewish state of Israel is politically in the ascendancy, supporting an exclusive form of Zionism which would make Jerusalem a predominantly Jewish city to the disadvantage of its other inhabitants and the Muslim sanctuaries there. From the perspective of traditional philosophy a lack of balance is apparent, and that deficiency is symptomized by the chronic unrest which marks modern Jerusalem as the point of confrontation between rival forces. In order to prove whether the philosophy inherent in these studies of the Heavenly Jerusalem is capable of serving any useful purpose, it has to be put to the test. And the most obviously suitable testing ground is Jerusalem on earth.

The inspiration for these final remarks is the thoughts and writings of Dr Yitzhak Khayutman, one of the planners of modern Israel and member of a group concerned with the nature and true meaning of Zionism. Zion is Jerusalem, and Zionism is therefore nothing more or less than Jerusalem-ism. It can thus be ascribed to the very people who proclaim themselves most hostile to it. The Palestinians who seek to regain and retain their share of Jerusalem are by that token Zionists; so are the Baptists and the many other Christian sects who call themselves Israelites and aspire to a place in the

Holy City; and even the fierce Iranians, whose declared intention is to march on Jerusalem and expel the Jews, have thereby adopted Zionism – in its exclusive aspect. It is that exclusiveness, affecting the extremists among Jews, Muslims and others, which constitutes the apparent obstacle to reconciliation. Yet it is also the extremists who make reconciliation possible. A tenet of traditional philosophy is that 'every action creates an equal and opposite reaction', and thus the two sides in any dispute, once they are clearly defined, make up an entity, like the two sides of a coin. Their interests are not merely rivals but also complement each other. To every reasonable person that is obvious. When all parties to the dispute over the earthly Jerusalem are brought to realize that each of them represents a form of Zionism, a reasonable accommodation between them is within reach.

Dr Khayutman's approach to the problem of pacifying the Middle East under conditions of lasting stability starts with his proposed redefinition of Zionism as 'the actualization of the Heavenly Jerusalem on earth'. That is certainly appropriate, for the primary function of the Heavenly Jerusalem and the studies associated with it is to identify opposites and to include them together in harmony within a conceptual framework which allows their differences to be transcended. In the same way as the New Jerusalem diagram provides the matrix which unites disparate systems of geometry, the concept of Heavenly Jerusalem is sufficiently wide to embrace all sides in any mundane dispute, set them together within an orderly pattern and thus demonstrate their essential unity. This method of reconciliation is based on transcendence, meaning that problems are raised to the point where the interests of all sides become identical. In the Middle East dispute the unifying element is the 'Jerusalem-ism' of all concerned parties. Their differences, expressed politically, seem irreconcilable, but on a higher level their aspirations are the same, to liberate Jerusalem and gain access to the spiritual powers which accumulate at the acknowledged cosmological centre of the earth. The fact that such powers are inexhaustible and can not be drained however much they are drawn upon suggests that in earthly Jerusalem, as in its heavenly original, there is room for all. In the plan of the New Jerusalem, together with the legends and traditions of the Holy City, are the principles for a scientific apportioning of Jerusalem's sanctuaries among all who have claims to them.

In the New Jerusalem diagram the twelve small circles on the periphery correspond in one of their aspects to the twelve tribes of Israel who were given possession of the Holy Land by divine covenant. They were traditionally placed in groups of three, north, south, east and west, and the pattern of their distribution was reflected in the design of the Tabernacle and the Temple which superseded it. That arrangement is repeated in the New Jerusalem diagram. The idea of applying it to the present-day situation at Jerusalem leads to an interesting question – where are the twelve tribes of Israel now to be discovered? The modern Jews claim to be made up of two tribes, Judah and Benjamin, together with an admixture of Levites. According to the Old Testament history (II Kings 17) all twelve tribes were removed from the Holy Land and taken captive into Assyria. Judah, Benjamin and some of the Levites have now returned to Jerusalem and hold power there. The other tribes are said to be scattered among the nations of the world. Attempts throughout history to identify them, or to claim identity with them, have involved some remarkable feats of imagination. Mystical writers have discovered them variously among the natives of North or South America, the Celts, Anglo-Saxons and all sorts of other peoples. As mentioned above, it is common among Christian sects to call themselves true Israelites and dream of entering the Holy City. But whoever and wherever the lost tribes may be, it is held alike by Jews and Christians that at some future time, immediately preceding a time of lasting peace, all twelve tribes will assemble again at Jerusalem. The prophecy is set out in, e.g., Ezekiel 37. Until such time there is no possibility of a peaceful settlement at Jerusalem.

The recognition of the missing tribes to make up the complete twelve is therefore of the greatest practical importance. And the most practical general principle, as it occurs to Dr Khayutman, is to allow the Israelites to declare themselves. That means to offer recognition as a legitimate Israelite to all who so call themselves or are impelled by Jerusalem-ism towards the Holy City.

A lasting form of settlement based on this ideal is possible, timely and, sooner or later, inevitable. It is in accordance with established prophecies, and there now appears in the form of the New Jerusalem the outline of the appropriate science for balancing forces, which is the instrument by which those prophecies may be fulfilled. That science is the justification for idealism, removing from it the stigma of impracticability. Its practical function, of course, is limited. It is one thing to clarify an ideal and another to live up to it. That difference is foremost in the mind of Plato in all his writings on idealism. The ideal city, he says, can never be reproduced on earth, because it is the nature of an ideal archetype to be mental and immaterial. On the other hand, without an ideal to aspire towards, society will always be unstable, reverting in the end to chaos. Plato's advice to the inhabitants of his 'best possible' state, Magnesia, was to contemplate the ideal cosmology – the pattern here referred to as the New Jerusalem – and copy it as far as possible in the design of every institution. After that it was to be a matter of compromise, of mitigating the ideal in the interests of practicality, adhering to it no more closely than circumstances permit. Thus if twelve tribes present themselves at Jerusalem, demanding recognition as true Israelites, the process of invoking the Heavenly Jerusalem can thereupon be initiated. If there be more or less than that number and if, as they will, they dispute each other's claims and credentials and engage in other forms of mischief, then the science of reconciliation through the New Jerusalem image will meet its expected test. Its hope of success lies in the fact that it is indeed a science, ideally adapted to present times and open to study by whoever cares to do so. Through that study are revealed the formulas which depict the structure of the universe in terms of its corresponding structure, number. Thence is developed a cosmological standard, a natural guide to civilized living. Sooner or later – and perhaps sooner than at present seems possible – the need for such a standard will become widely apparent. Despair at the impotence of political diplomacy will make plain the necessity for transcendence, and minds will be directed again into the channels of traditional thought which converge upon the vision of the New Jerusalem.

The main text of this book has been the twenty-first chapter of St John's Revelation with its description of the Heavenly Jerusalem coming down to earth. In the chapter which follows, the last chapter in the Christian Bible, the New Jerusalem is again referred to, its twelve 'lunar' circles representing the twelve fruits of the Tree of Life:

> . . . and on either side of the river was there the tree of life, which bare twelve manner of fruits, and yielded her fruit every month: and the leaves of the tree were for the healing of nations.

The leaves, $\phi\upsilon\lambda\lambda\alpha$, have the same number as the New Jerusalem, 961.

The City in Greek idealistic philosophy and the Tree of Life in Jewish mysticism are images of the same archetype, the ideal cosmology, a function of which is 'the healing of nations'. St John's prophecy of the sacred cosmology restored to human consciousness through the image of an all-inclusive City or a healing Tree is of particular significance to Christians because it occurs at the very end of the Scriptures. Even those people with minds inimical to prophecy may acknowledge the modern relevance of St John's imagery. His ideal City, to which are drawn all 'the nations of them which are saved' under the rule of spirit, previses a necessary and therefore inevitable state of affairs to come. That state, a product of minds reformed by necessity, will be of structured harmony, the key to which is in the scientific code described in this book and symbolized by the Heavenly Jerusalem.

Index of numbers

The numbers discussed in the text are indexed not in order of magnitude but on a principle like alphabetical order. Thus 2187 precedes 220. Decimal divisions and multiplications of a number, such as 10.8, 108, 1080, 10800, etc., are generally listed under one heading.

Ken
Vesica,
gematria
coeval

_____